D0904889

WITHDRAWN

TOURO COLLEGE LIBRARY
Women's Building

Polyphonic Minds

WITHDRAWN

TOURO COLLEGE LIBRARY
Women's Building

Polyphonic Minds

Music of the Hemispheres

Peter Pesic

The MIT Press
Cambridge, Massachusetts
London, England

3360

ωB

© 2017 Massachusetts Institute of Technology

All rights reserved. No part of this book may be reproduced in any form by any electronic or mechanical means (including photocopying, recording, or information storage and retrieval) without permission in writing from the publisher.

This book was set in Syntax LT Std and Times New Roman by Toppan Best-set Premedia Limited. Printed and bound in the United States of America.

Library of Congress Cataloging-in-Publication Data is available.

ISBN: 978-0-262-03691-7

10 9 8 7 6 5 4 3 2 1

3/14/22

for

my polyphonic minds

Contents

IV   POLYPHONIC BRAINS   227

Prelude

Western music embraced polyphony, the simultaneous interweaving of several different musical lines or voices. Yet, as familiar as it may seem, the practice of hearing or thinking many things at once deserves closer scrutiny. This book examines the nature and larger significance of "polyphonicity," of "many-voicedness" both in music and human experience as a whole. To do so requires examining what polyphony has meant over the long span of time between antiquity and the present, during which it went from being scarcely noticed to defining the apex of musical art. Western polyphony emerged in a charged relation with Christian theology but has become ubiquitous: beginning as a music suitable for divine minds, for gods and angels, in time polyphony became a defining attribute of human personhood.

The ensuing story in many ways encompasses the history of music but also involves even larger questions: How can a single mind apprehend several different things simultaneously? What allows that mind to experience those things as a unity, rather than an incoherent jumble? We will address this question from many angles, as is appropriate for an effort to understand how unity could arise from multiplicity. I hope to show the continuing fascination of this question, as well as consider the many responses it has evoked. In an earlier book, I explored the ways in which music influenced the making of modern physical and mathematical science.[1] This book will argue that polyphonic music raised fundamental issues for theology, philosophy, literature, sociology, psychology, and neuroscience, all searching for the apparent unity of consciousness in the midst of multiple simultaneous experiences. Accordingly, this book will range through all these domains as it presents the development of polyphonic music and the significance of polyphonicity for the human sciences as well as for the arts.

Ordinarily, we seem to attend to one thing at a time, for even a person who is "multitasking" usually does not work on all those tasks at once but turns from one to another as needed, really engaging in something more like "task switching." Saying that "something else was on my mind" implies that I was not really paying attention to the matter at hand. Then too, when surrounded by several nearby conversations, equally loud (say at a party), I cannot understand them all but may try to direct my attention from one to another, though only with difficulty: their very multiplicity distracts. This "cocktail party problem" raises classic questions in cognitive psychology, especially to what extent can the brain process multiple simultaneous inputs.[2] We shall return to this problem at several points.

Though initially posed in laboratory experiments (such as listening to two different streams of words through each ear), this problem now has grown to a near-universal dilemma. Many people now multitask constantly, checking their smartphones, texting, and using social media while trying to do their work or have social interactions; 9 percent of adults report using their phone during sex.[3] This contemporary problem of the "distracted mind" raises the specter of growing inability to concentrate or focus, not to speak of the immediate danger of multitasking: 23 percent of all car crashes involve cell phone use.[4]

In contrast to modern multitasking, Aristotle taught that the mind cannot think many things at once, arguing that "it is possible to know many things but not to be thinking of them," here distinguishing *potential* subjects of thought, which may be multiple, from the one thing I am *actually* thinking right now.[5] Aristotle's assertion might help explain why we struggle to understand multiple spoken conversations, but it leaves untouched the problem of how we can make sense of several simultaneous musical voices (or "lines"). In contrast, his teacher Plato had argued that the psyche was composed of three parts (specifically those by which one learns, is angry or combative, and by which one desires), hence raising the possibility of their disunity.[6] During the twenty years they spent together, he and Aristotle must have often argued over whether the mind is fundamentally one or many. If each individual human mind is essentially *one*, how can it deal with the multiplicity of what it perceives? If each mind is not a single thing, how can we seem to understand or grasp anything *as one*? How could we even form the concept of unity, much less experience it in the welter of multiple voices, whether inner or outer?

This book emerged from my struggle with this question, which I encountered in various forms in music, philosophy, and psychology. Nonetheless, I will argue that music is not only a striking instance of this general problem of the Many and the One but may reveal its crux. The history of musical polyphony opens a window on its implications, both in situations where polyphony was avoided or ignored and in those where it was cultivated.

Exploring the ramifications of polyphony initially involved theological and philosophical questions that deeply engaged medieval thinkers. Yet the relation between the Many and the One impinges on ultimate concerns that go beyond any one religious tradition, even though they were initially phrased in terms of certain highly controversial questions of Christian theology. In our time, the problems of polyphony reflect neuroscience's quest to understand how the many simultaneous processes in the human brain lead to the appearance of a unified consciousness.

The musical form of our problem concerns the possibility of hearing or playing many different, independent musical lines at once. Here, I will use the common term "music" to include many kinds of sonic experience, some of which will not fit in the ordinary connotations of the "musical." Indeed, I do not assume that readers are necessarily familiar with the needed musical terms and concepts but will try to explain and explore them from the ground up. To ensure a common frame of reference, I will generally use modern terms, some of which are in wide use, others of which have to be devised anew. The very difficulties that emerge in framing these terms will raise important and helpful philosophical questions. Along the way, I will also try to highlight the ways these terms have changed in meaning and usage, for that history illuminates crucial changes of understanding.

At one extreme, *monophony* or *monophonic music* consists of a single melodic line without accompaniment or additional parts, such as Gregorian chant or many kinds of traditional melody throughout world music. This is a relatively recent word compared to original terms such as *song* (*carmen* in Latin) or *plainchant* (*cantus planus*). At the other extreme, I will call *multiphony* an experience of many different sounds without any ongoing coordination or imposed coherence, for example listening at a party to several nearby conversations. I find this term preferable to the pejorative *cacophony* (literally, "bad sounds"), because multiphony need not be unpleasant, as we shall see through the work of John Cage.

Less extreme than multiphony, *polyphony* or *polyphonic music* combines several musical lines, each of which retains its identity as a line to some degree. Though the term "polyphony" first appeared in thirteenth-century texts, it only entered the English language in the eighteenth century; as this book unfolds, we will encounter such original terms (each with a different specific meaning) as *organum, discant, figured song, counterpoint,* and *pricksong*.[7] Beginning in the ninth century CE, Western art music contains many examples of polyphony, from medieval organum and motets to Renaissance works for two to forty individual melodic lines (usually called *voices*, even though each such line might be performed by one or many singers or instruments) by composers such as Josquin des Prez or Giovanni Palestrina, through Baroque compositions (especially by J. S. Bach), and many works since then, until the present day.

Moving closer to monophony, *homophony* has come to mean music in which one voice leads melodically, while the other voices may follow its rhythms or may accompany it in ways that indicate their dependence on it, as opposed to the greater independence of the voices in the more strict forms of polyphony we will consider. Examples of homophony include barbershop quartets or hymn singing. Likewise, *heterophony* is an improvised kind of polyphony in which different performers use more or less modified forms of the same melody; in many kinds of traditional music, different singers might add extra tones or ornaments to a common basic melody, so that the effect is somewhere between pure monophony or homophony and polyphony. In 1928, traveling in central Africa, André Gide described the heterophony of the Sara people, an ethnic group he encountered there: "Imagine this tune yelled by a hundred persons, *not one of whom sings the exact note*. It is like trying to make out the main line from a host of tiny strokes. The effect is prodigious, and gives an impression of polyphony and of harmonic richness."[8]

These various kinds of music involve different degrees of multiplicity of voices, each of which may be more or less strongly related to others being sung at the same time. Though I will try to use the appropriate terminology from the period in question, at times I will (to be clearer) use terms that may have come into use long after, drawing attention to such anachronisms so that they are less misleading. In general, I will use "polyphony" to encompass a wide variety of simultaneous sounds, which may be more or less independent or connected.[9] When it is important to draw attention to their simultaneous interrelation at a certain moment, I will use the common terms *harmony* or *chord* (though these only came into use in recent centuries). If we think of time flowing along a horizontal axis (as in a musical score), we can see why these have also come to be called "vertical relationships" of simultaneous sounds. Conversely, when we need to emphasize that different sounds are part of separate melodic voices, I will speak of *counterpoint* and the "horizontal relationships" within each of the constituent "melodic lines." These terms span a nuanced spectrum containing many possibilities of multiplicity or unity; we will need to use them flexibly as tools to describe several musical dimensions. The diverse kinds of music they describe will offer us a rich field of possibilities, which will both raise and help us address the problem of the Many and the One.

As we assess these developments, it will prove important that Western music is not a single tradition. Most listeners are familiar with *music as expression* evoking various passions, which Vincenzo Galilei (father of Galileo) called "ancient music" (*musica antica*).[10] We will also need to attend to an important alternative tradition, *music as science*, often associated with cosmic harmonies or the divine, dispassionately transcending human emotion. Beginning in ancient Greece, this dispassionate musical science stood with its sisters—arithmetic, geometry, and astronomy—in the *quadrivium*, the "four-fold way" at the center of "liberal education."[11] Particularly in relation to polyphony, Vincenzo called this "modern music" (*musica moderna*).

These two different musics, the expressive and the scientific, have coexisted since ancient times.[12] Their complex relation has shifted over time; though the dispassionate musical science enjoyed greater prestige in ancient and early Christian times, by the sixteenth century the balance had shifted. Indeed, Vincenzo argued that a revived expressive music should eclipse modern polyphony in order to create a newly powerful dramatic art. Around 1600, the first operas gave practical form to his quest to retrieve the lost powers of ancient drama. This represented a musical revolution that preceded the scientific revolutions and led to the predominance of expressive power as the goal of music, rather than beauty, which was the goal of music as science and dispassionate practice. Polyphony then confronted newly heightened dramatic declamation: How could the interweaving of many voices deal with the expression of a single feeling?

As a teenager, I became fascinated with scores of modern music, such as Stravinsky's *Rite of Spring* (figure 0.1; ♪ sound example 0.1). Its thirty-three vertically aligned staves contain directions for a large symphony orchestra of about a hundred players. The moment shown is particularly frenetic; a mysterious sage has arrived to officiate at a ritual in which a chosen maiden dances herself to death, part of fertility rituals in ancient pagan Russia. All the instruments are playing fortissimo or even fortississimo: literally, "as loud as possible" and "even louder than possible." Many different rhythms and timbres are sounded insistently, especially by the percussion. But as amazing as this theatrical moment might be, I wondered how anyone could register all these simultaneous sounds and rhythms—especially the composer or the conductor. I struggled not to lose my place in the welter of sound as I listened to a recording and tried to follow the score. My eyes, ears, and mind all tried to find their way in this buzzing hive of sound.

In a way, Stravinsky's score asks our question from its very beginning. Its first measures (figure 0.2, ♪ sound example 0.2) begin with a single instrument—monophony—a solo bassoon playing in its highest register, so unusual a sound in its time that some members of its first audience jeered or bleated to mock what struck them as weird, even comic. Then a horn enters (measure 2), but seems to accompany the bassoon. Is this homophony, or is it really polyphony? Clarinets enter two measures later; the musical texture becomes more and more polyphonic, leading eventually to our first example (figure 0.1), bristling with different voices whose interrelation is unclear, perhaps even antagonistic. At the end of this example (rehearsal number 71 in the score) all one hundred instruments suddenly fall silent: a "general pause" of one measure ensues, the fermata ⌒ indicating that the conductor may extend it at will, the marking *lunga* ("long") indicating a prolonged dramatic silence. But in the wake of what has just happened— a kind of ever-intensifying racket, an angry crowd of instruments—even this "silence" seems unquiet, filled with foreboding: the musical texture has been sheared off abruptly, halting as mysteriously as it began. Perhaps a "silence" cutting short so many voices means something different than one that might interrupt a single melodic line. The wild dancing that ensues after this point in the score richly responds to that charged pause.

Turning back to the beginning of the work, I thought perhaps I should follow one line of music, one single "voice." But the moment a second instrument entered, I was not sure where to look. My eyes darted back and forth between the two lines of score, mainly following what seemed to be the predominant voice, at first the bassoon, but later other instruments. I often gravitated toward the staff for the instrument I played since childhood, the violin, whose sound was familiar. I wondered whether I was really reading the "whole" score, or even two voices at once, rather than just becoming faster and more adept at glancing back and forth between staves, though actually never taking in more than one of them at once.

This difficulty was natural for someone who played an instrument like the violin, notated on a single staff. But at that time I was also beginning to play the piano, struggling to play two lines at once, one in each hand—and each line could contain many notes. How to combine them? I would learn the hands separately (struggling with the unfamiliar bass clef), then try to put them together. At first this was frustratingly difficult, even impossible.

Figure 0.1
Igor Stravinsky, *Le sacre du printemps* (1913) in orchestral full score, the end of "Cortège du Sage," ending at rehearsal number 71 with a "general pause" (G. P.) (♪ sound example 0.1).

Figure 0.2
Igor Stravinsky, *Le sacre du printemps* (1913), the beginning of the introduction (♪ sound example 0.2).

How could I do two or more different things at the same time? But gradually I started to be able to do it—or at least both my hands were playing the requisite notes at the indicated time, if only in fits and starts. Even when I succeeded, I was mystified: how had I done it? All I really felt I understood were the tasks of the two separate hands, but when I put them together I ceased to understand and just surrendered to some kind of automatism that my hands were able to execute, somehow—each playing its appointed notes—which I could not really comprehend, only play. Fifty years later I still wonder.

These issues involve deep problems that span music and mind. I have tried to offer a synthesis that breaks new ground by placing polyphony in the widest context of musical, philosophical, theological, and scientific concerns, highlighting the questions surrounding polyphony and engaging those questions anew. Though I am not aware of any earlier work that has tried to do this, at least so widely, I am indebted to many specialized studies, for a synthetic work like this necessarily goes beyond the material familiar to any single person. I have been helped by the work of Thomas J. Mathiesen, James Haar, and Blair Sullivan on ancient and medieval theory; Craig Wright on Notre Dame; Francesco Ciabattoni on Dante; Patrick Macey on Savonarola; Rob C. Wegman on the crisis of polyphonic music; and Anne Harrington on the history of neuroscience, to name only a few.[13] The notes and boxes contain additional details for those who wish to go further.

Throughout, I will blend historical consideration with philosophical reflection because both are important to my overall inquiry, which is situated at the borderland between practice and theory. To be sure, the history of polyphony has been well studied within the internal history of musicology, whose findings I hope to bring to wider attention as well as add something new. My account of that history does not pretend to be complete but instead tries to frame some of its important episodes in a new way within the larger philosophical and theological issues that polyphonicity raises. To illustrate these issues, I have often chosen rather familiar musical examples, many of them staples of music history texts and anthologies. This makes it easier for readers to find and explore more performances of the works considered. I have sought to say something helpful to those hearing these works for the first time, while also offering a new perspective to those for whom they are familiar.

This book's four parts begin with the emergence of polyphony and its flowering, then turn to the horizons of polyphony and our own polyphonic brains. Chapter 1 begins by situating polyphony in a global context, though thereafter I will restrict myself to its history in the West. In ancient Greece, Plato gave the earliest recorded critique of the power of music to move the passions through monophonic songs, while Aristotle defended the positive effects of the catharsis such music could induce. Though Plato criticized the use of polyphony in ordinary singing, he presented the "music of the spheres" as a harmony of many voices, giving the earliest recorded presentation of polyphony as a significant phenomenon. In Roman times, Plato's daring idea found resonance in Cicero's musical evocation of an ideal commonwealth and later in Martianus Capella's Neoplatonic vision of the unfamiliar polyphonic music that accompanies the highest divine intelligence.

Turning away from the theatricality of pagan music, early Christian practice chose dispassionate prayer and hymnody as most appropriate for the exalted godhead, following the ancient tradition of musical science. To reach the divine, passions should be purged and purified, not encouraged, as the Greek father Evagrius of Pontus argued. Polyphony entered the formal practice of Western music during the ninth century as part of Christian liturgy, which previously had relied on monophonic chant. This momentous change raised many theological and philosophical issues, as I relate in chapter 2. Difficult and highly contested concepts of the Trinity, of the relation between the divine and the human, and of the meaning of personhood parallel musical questions about the relation between the individual voices and the polyphonic whole. Medieval Neoplatonists such as John Scotus Eriugena presented polyphony as reflecting the outpouring of the divine One into many voices. Eriugena went too far for the Western church, which ultimately rejected his radical vision that all human beings, not just the Christ, could find union with God. Western orthodoxy also condemned him because rebellious groups such as the Albigensians rallied around his books. These controversies reflected harsh political realities about the predominance of the church and the emergent French monarchy in the face of what they deemed heresy and rebellion.

Nonetheless, polyphony continued to grow in complexity, especially around the new cathedral of Notre Dame, in the midst of the philosophical and theological controversies that galvanized scholars and students in Paris. Chapter 3 presents the powerful advocates as well as critics of the new polyphony. Though Thomas Aquinas argued that God and the angels could know many things at once, he agreed with Aristotle that human beings were limited to knowing only one thing at a time. Thomas thus took the orthodox position against radical Neoplatonic views about the divinization of human beings. As a Dominican friar, he agreed with the teaching of his order that music should be "simple," meaning monophonic chant. In contrast, though he admired Thomas, Dante praised polyphony as the appropriate music for Paradise and made explicit what he considered the "personhood" of a polyphonic work. For Dante, the union of many voices reflected and glorified the communion of the blessed souls.

Though elaborate polyphony was formally prohibited by papal edict, it continued to flourish. For instance, Nicole Oresme, the most important natural philosopher and mathematician of the fourteenth century, was friends with Philippe de Vitry, the preeminent composer of the polyphonic "new art" (*ars nova*), as I recount in chapter 4. Oresme thought that such polyphony exemplified the "new song" praised in the Bible. Polyphony also connects with his investigations of cosmic recurrences as well as of new mathematical ways of representing physical quantities using several "dimensions" (as he called them).

Controversies about polyphony continued to grow in the late fifteenth century, bringing forward long-simmering issues concerning the use of complex artifice in church music, increasingly contested as the movement for reformation grew toward its crisis, even leading at one point to armed combat between impassioned partisans. Chapter 5 outlines the polemic between such reformist figures as John Wyclif, Girolamo Savonarola, and Desiderius Erasmus, who attacked polyphony, and others who rose to its defense, including Martin Luther. Leonardo da Vinci argued that polyphony gave music its true intellectual and artistic significance, able to excel poetry and even rival painting.

Polyphonic compositions of the Middle Ages and the Renaissance used several techniques to unify their several voices, as I describe in chapter 6. Works such as Guillaume de Machaut's *Messe de Nostre Dame* took up techniques already in use by his Parisian predecessors, building the voices around a single melody (the *cantus firmus*), often a Gregorian chant, but shaping them further through *isorhythm*. *Canon* or *imitation* makes all the voices use the same melody more strictly, as we will see in works by Johannes Ockeghem and Josquin des Prez. When composers used very large numbers of voices, they simplified their harmonies because the voices have to share notes in order to avoid excessive dissonance, resulting in newly unified textures that subsumed the interplay of many voices. As chapter 7 details, motets in forty parts by Alessandro Striggio and Thomas Tallis so impressed their contemporaries that these works figured in dynastic struggles and international politics.

Whatever the number of voices, the character of a polyphonic composition depends crucially on how it uses and resolves dissonances, treated in chapter 8. The suave serenity of Giovanni da Palestrina's music depends on his exacting control over dissonance, its careful preparation and resolution, while Carlo Gesualdo draws expressive force from unprepared dissonance. After 1600, these expressive devices became crucial to the "second practice" of Claudio Monteverdi and others that led to the new dramatic genre of opera. Around the same time, the practices of counterpoint became codified in rules so that musicians (including a large body of serious amateurs) could learn and apply them to performance and composition. Didactic works by Gioseffo Zarlino, Thomas Morley, and Johann Joseph Fux moved from teaching the precepts of contrapuntal practice to outlining the rules of a new phase of musical science, as chapter 9 describes. J. S. Bach considered himself a practitioner of that science and devoted much attention to its teaching, particularly through the practical examples he provided in his musical works. Further, his innovations in the digital techniques of keyboard playing deeply changed the practice of counterpoint. Chapter 10 argues that his new use of the thumb amounted to a new technology that enabled a new level of keyboard counterpoint exemplified in his fugues. Further, he pioneered a "virtual polyphony" in which a single instrumental line could suggest the interplay of several voices. For his contemporaries, Bach's fugues represented a rhetorical achievement that was tantamount to the creation of an artificial person, a work of art that enacted idealized persuasion, uniting music as science and as expression.

In the centuries following Bach, composers created new hybrids of polyphony and homophony. Chapter 11 illustrates various possibilities of polyphonic complexity and its artistic effects in the compositions of Ludwig van Beethoven, Johannes Brahms, and Arnold Schoenberg, including new ranges and applications of orchestral timbre that reflected many technological innovations of instrumental practice. Through its own innate dynamic, music as a project to ever intensify expressivity approached a self-consuming crisis. These developments overlapped with the physiological studies of Hermann von Helmholtz, who explained dissonance in terms of the degree of clashing between overtones. Helmholtz also emphasized that even a single tone contains a multiplicity of overtones, whose relative strengths explained its particular timbre (sound color). At the same time, the explorations of orchestral color by Schoenberg and others led to further expansion of polyphonic possibilities: simultaneous different rhythms, even tempos, were further exploited by Charles Ives, Karlheinz Stockhausen, György Ligeti, and Conlon Nancarrow.

The meaning of polyphony may be further extended to include "voices" that go well beyond the traditional limits of music. Chapter 12 begins with the contrapuntal radio compositions of Glenn Gould, which explored the multiphony of several simultaneous speakers. Going further still, John Cage enlarged the domain and meaning of polyphony to include a continuum of "voices" that transformed the meaning of sounds and silence. The concept of "field" provides a helpful generalization of distinctly separate voices to a fluid continuum in which an indefinite number of elements mingle freely while preserving the fundamental tensions in the ways they coexist and interact.

The general concept of polyphonicity also encompasses new ways of discussing literature, psychology, and sociology, presented in chapter 13. In psychology, the "voices" within the human psyche could become so dissonant as to verge on being a collection of disunified selves. Such concerns grew with the development of neurology and the growing awareness of the specialized roles of the cerebral hemispheres, increasingly connected with clinical cases of "multiple personalities." These investigations also included musical behavior, especially the polyphonic activity of piano playing. As he shaped the new discipline of sociology, Max Weber presented polyphony as a central factor in the "rationalization" of Western music and society. In literature, Marcel Proust presented the inner drama of the many selves that live within each character. Amy Lowell and Mikhail Bakhtin each enlisted polyphony to express their new visions of poetry and narrative. Leon Festinger coined "cognitive dissonance" to clarify the felt inner conflicts that shape social psychology; Pierre Bourdieu and Bernard Lahire disagreed about the degree of coherence in the multiple aspects of each person's identity and social activity.

The brain, as currently understood, operates through simultaneous processing involving many internal subcenters. Chapter 14 reviews the development of the neuronal approach to brain function as well as the relation between "split brain" studies of the roles of the hemispheres and "field theories" that emphasize the plasticity of the brain, rather than localized cerebral functions. The interconnected activity of these cerebral subcenters embodies both sides of polyphony, the multiplicity of simultaneous voices and the changing dissonances by which they mutually interact. The increase of audible dissonance in the neighborhood of consonances may illuminate the characteristic "U-shaped curve" in learning a complex skill, the troubling phase in which greater effort seems to yield less success. Indeed, the distractibility of attention-deficit/hyperactivity disorder (ADHD) seems connected with a particular mode of experiencing the polyphony within the brain.

The study of brain rhythms remains an important stream of twenty-first century neuroscience. Chapter 15 considers the status of this project as presented by the neuroscientist György Buzsáki. His laboratory and others have undertaken an extensive investigation of the interacting neuronal rhythms that polyphonically constitute the "music of the hemispheres," which shape brain states from sleep to awakening. Neuroscientific discourse about the exact characteristics of the "neural orchestra" uses concepts drawn from polyphony to give new ways to describe the functioning of the brain, ways difficult for ordinary language but natural for music. Thus, polyphony, as practice and concept, remains a challenge for neuroscience, particularly the ways the brain processes polyphony, which thus far remain underexplored, compared to other investigations of the relations of music to consciousness.

Our story begins with the quest to connect polyphony with divine or angelic minds and ends with human brains. Yet we will have to go beyond the realm of purely mental constructs. The human experience in all its polyphonicity involves rhythm in ways absolutely dependent on the body. Our "neural orchestra" cannot be understood without the dance of the body that sustains the brain and enables the full activity of our polyphonic persons.

Throughout the book, when I refer to various "♪ sound examples," please see https://mitpress .mit.edu/books/polyphonic-minds (sound examples can be accessed in most web browsers). See that link as well for further information on purchasing enhanced digital editions that will be available in a variety of formats. The text and examples are most easily and seamlessly available in the e-book, in which you need merely touch an example to hear it.

POLYPHONY EMERGENT

1 Global Contexts and Ancient Origins

Though we will restrict ourselves mostly to the Western musical tradition, we begin by clarifying the larger context of monophony and polyphony, which are found throughout the world, distributed through time and space in ways still not fully understood. The origins of Western polyphony come from ancient Greece, whose music seems to have been largely monophonic, though with some evidence of polyphonic practices. Plato recognized the emotional power of monophony and critiqued contemporary attempts to add additional notes to a single melodic line. Yet he also described the exalted, impassive "music of the spheres" as simultaneously sounding notes. His visionary account resonated with Roman authors such as Cicero and Martianus Capella, who described a "profound ancestral song" of polyphony flowing from the divine unity.

Until the late nineteenth century, Western scholars thought European polyphony was completely unique.[1] They assumed that the rest of the world's music was monophonic, comparable to the European folk songs they knew. That view changed radically through better acquaintance with the breadth and variety of musical practice made possible by careful fieldwork.[2] The Western study of music tended, until relatively recently, to concentrate on its own European traditions from the Middle Ages on, as well as those ancient civilizations that were considered direct ancestors. More wide-ranging study of world music began in earnest during the nineteenth century, starting with great empires such as India and China, and later going further afield.

Ethnomusicologists now regard polyphony not as exceptional but as almost ubiquitous. The map in figure 1.1 shows the distribution of primarily monophonic versus polyphonic cultures, as they are now known. This map raises many unsolved questions: why are North and South America, along with East and North Asia, so predominantly monophonic, in comparison with the many polyphonic traditions of sub-Saharan Africa or Eastern Europe? What explains isolated "islands" of polyphony (like the Ainu of Japan or the Andean Q'ero) in the middle of "oceans" of monophony?[3]

We cannot assume that monophony and polyphony represent two simple, exclusive alternatives, nor that those modern Western terms are fully appropriate to describe all the possibilities. A wide spectrum lies between monophonic chant and a wild jam session in

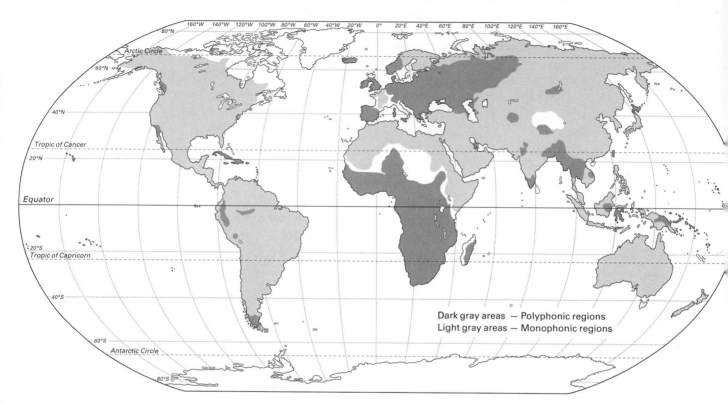

Figure 1.1
Map showing distribution of primarily monophonic (light shading) versus polyphonic cultures (dark shading).

which a casual listener can scarcely discern any coordination between the voices, not to speak of the multiphony of cocktail party conversations mentioned above. Compared to these extremes, the middle ground of mild heterophony may be more comfortable and more widespread. In many musical cultures, singers do not insist on exact unison but join the ensemble more freely, not necessarily diverging from the other voices but perhaps starting a bit late or varying the melodic line a little as they join in.

Besides the common use of such informal heterophony, Georgian traditional music (for instance) has a well-developed tradition of more elaborate heterophony and complex polyphony dating back well before Christian times. The example in figure 1.2 (♪ sound example 1.1) illustrates one of the most complex forms, from Guria in western Georgia.

Then too, many musical traditions in Africa include complex polyrhythms, far more intricate than common European traditional practice, often using percussion instruments (such as figure 1.3; ♪ sound example 1.2).[4] Such examples, among many others, have left little support for earlier views of "primitive" versus "high" art, or for simple presumptions about "natural" musical style. The Banda-Linda horn orchestra of the Central African

Figure 1.2

"Adila," a Gurian trio song (first half) from western Georgia, transcribed by Nino Tsitsishvili (♪ sound example 1.1). Text: "A rose was asked: Why are you so pretty, your face and body? I am surprised that you are so thorny and so hard to find."

Figure 1.3
"Ndrodje balendoro," a Banda-Linda orchestral piece from the Central African Pygmy people, transcribed by Simha Arom (♪ sound example 1.2).

Pygmy people involves eighteen different parts interlocking in a complex "hocket" (as this technique was called in medieval Europe), the rapid alternation of several voices, with one sounding while another rests.[5] As John Michael Chernoff observed, "only through the combined rhythms does the music emerge, and the only way to hear the music properly … is to *listen to at least two rhythms at once*."[6]

Such polyphonic practices parallel the social and political organizations of which its musical life is an integral part. In the egalitarian society of the Pygmies, anthropologists note that it is rude to tell someone else what to do: men cannot order around their wives nor parents their children, nor does social status confer authority. Still, "the camp spontaneously organizes itself to find sufficient food without an elder or leader directing people to act. People organize themselves sensitively in relation to what others announce they are doing, so that their actions are complementary."[7] From infancy on, BaYaka Pygmies are regularly immersed in performing their complex polyphonic music, which instills, reflects, and reinforces their social patterns. As the anthropologist Jerome Lewis observed of this group,

participating appropriately in a song composed of different parts sung by different people simultaneously involves musical, political, psychological, and economic training. Anyone can start or stop a song, though there are particular conventions to follow. There is no hierarchy among singers, no authority organizing participation; all must be present and give of their best. All must share whatever they have. Each singer must harmonize with others but avoid singing the same melody; if too many sing the same part, the polyphony dissolves. Thus each singer has to hold their own and resist being entrained into the melodies being sung around them. Learning to do this when singing cultivates a particular sense of personal autonomy: one that is not selfish or self-obsessed, but is keenly aware of what others are doing and seeks to complement this by doing something different.[8]

Colin Turnbull noted that "the chief delight of the singers is to listen to the effects of group-produced counterpoint as it echoes through the dark cathedral of the jungle."[9] We will return repeatedly to this implicit contrast between such a polyphonic society, cultivating the interplay of diverse voices, and a monophonic one that strives to unite all the voices into a single whole reflecting the will of a supreme ruler or deity. Indeed, Alan Lomax argued that older egalitarian cultures, characterized by polyphonic choral singing, over time were replaced by dictatorial societies that sang in complex monophony.[10]

Similarly, Joseph Jordania has suggested that polyphony was once widespread, if not universal, though subsequently many societies abandoned it or relegated it to an isolated part of their repertoires.[11] According to some scholars, the oldest known music may indeed have been polyphonic. Dated about 1200 BCE, a cuneiform tablet from the Babylonian city of Ugarit (near Ras-Shamra in modern Syria) contains a hymn in the Hurrian language to the goddess Nikkal, wife of the moon god. As deciphered in 1976, this hymn presents a solo voice accompanied in the same rhythm by a harp or lyre that does not simply duplicate the sung pitches but adds an additional pitch.[12]

This connection to a female deity may reflect larger cultural factors: Lomax argued that, "since female voices, the world over, operate about an octave higher than male voices, I take the presence of two parts to be a simultaneous communication of male and female roles; and the effect is the same whether the chorus is all-male or all-female … [so that] the upper voice part stands for the feminine." From his worldwide data, Lomax concluded that counterpoint ("two or more parts which are rhythmically and melodically independent") "turns out to be most frequent among simple producers, especially among gatherers, where women supply the bulk of the food. Counterpoint and perhaps even polyphony may then be very old feminine inventions."[13] Likewise, in early hoe agriculture, women take a leading role and "are not so likely to be shut away from the public center of life. … It is in such societies that we find the highest occurrence of polyphonic singing," reflecting the higher "level of complementarity" between the sexes: "Vocal polyphony, then, may be viewed as a communication about the recognized importance of the feminine role in culture."[14] Conversely, "among Amerindian hunters and fishers, where most performances are dominated by males, vocal unison by far outweighs all other performance modes and polyphony scarcely exists."[15]

Despite the number of polyphonic musical cultures, much world music is basically monophonic, based on a single melodic line, more or less strictly maintained. In particular, the musical traditions of China and India did not rely on multiple simultaneous melodies.[16] The Western emphasis on polyphony remains noteworthy for its insistence on particular kinds of independence between voices that are also correlated in ways that we will later examine in detail. Though the pre-nineteenth-century insistence on the uniqueness of Western polyphony may have been overstated, ancient Greek musical practice (as currently understood) does seem to have been largely monophonic.[17] Here, we are handicapped by the paucity of extant notated pieces of ancient Greek music, hence are forced to rely on the surviving texts that describe them and their theoretical basis.[18] It is as if we had no texts of the Greek tragedies, only descriptions of their effects; indeed, since those tragedies were certainly danced and sung, the texts that remain essentially give us only their librettos, so to speak, lacking a musical notation that would enable us to gauge their full effect. Still, the few fragments of Greek music that have survived provide important clues.

The extant works of Greek musical theory concentrate on a single melodic voice. Plato treated music as a central aspect of education and philosophy and devoted special attention to melody (*melos*). In Plato's *Republic*, Socrates excludes all melodic modes that stimulate "drunkenness, softness, and idleness," leaving only "the mode that would suitably imitate the tone and rhythm of a courageous person who is active in battle or doing other violent deeds." Besides this, he only allows another mode for "someone engaged in a peaceful, unforced, voluntary action," such as praying to a god, persuading another person, instructing or being instructed, "not with arrogance but with understanding."[19] Thus, Plato acknowledged the long tradition of *musica antica*, as Vincenzo Galilei called the wide repertoire of songs that stimulate passions of all kinds. Plato considered their power and importance so great that he argued for eliminating all melodies other than these two positive kinds, famously inviting the tragic poets to leave his ideal city in order to rid it from their weeping and wailing. In contrast, Aristotle argued that "all the modes must be employed by us, but not all of them in the same manner. … Those who are influenced by pity and fear, and every emotional nature … are in a manner purged and their souls lightened and delighted. The melodies which purge the passions likewise give an innocent pleasure to mankind."[20] Using such melodies, tragic drama induces purgation or purification (*katharsis*) through the "tragic pleasure" of pity and fear.[21]

These celebrated treatments of music centered on melody, its modes and rhythms. If the ancient Greeks did add accompaniments, they scarcely commented on them, by comparison with their extensive discussions of the melodic modes. Yet in Plato's *Laws*, the Athenian Stranger remarks that the lyre "must produce notes that are identical in pitch to the words being sung," implying that singers would accompany themselves by playing on the lyre, if only to give themselves their starting pitch and support their melodic line.[22] Such practices of accompaniment (now commonly called *heterophony*) have been observed in Ethiopian singing to the lyre, as well as in many other traditions involving the performing needs of a singer-composer, including the Homeric bards who sang to the lyre-like phorminx.[23]

Though aware of this, the Athenian Stranger nevertheless goes on to argue that

the lyre should not be used to play an elaborate independent melody [*heterophonia*]: that is, its strings must produce no notes except those of the composer of the melody being played; small intervals should not be combined with large, nor quick tempo with slow, nor low notes with high. Similarly, the rhythms of the music of the lyre must not be tricked out with all sorts of frills and adornments. All this sort of thing must be kept from students who are going to acquire a working knowledge of music in three years, without wasting time. Such conflict and confusion makes learning difficult.[24]

This critique implies that such things did indeed happen frequently enough to annoy Plato's Stranger, though he does not disclaim what he considers the proper or seemly use of the lyre accompanying the voice.[25] Viewed in the larger context of the musical discussions in Plato's *Laws*, such elaborate and increasingly complex polyphonic developments seem to have been part of newer musical currents that disturbed his conservative sense of received musical practice.[26]

For Plato, such new developments had great political import because he judged they would lead to an increasingly individualistic, fragmented polity. The Stranger regrets the divergence between accompaniment and song, whose "conflict and confusion" distract learners. Virtuoso performers would be tempted to show off their skills through *heterophonia*, a word Plato seems to have used to denote departures from strict unison music, as compared with the modern usage of "heterophony" to denote the simultaneous use of different forms of the same melody in different voices, as when an accompaniment adds extra tones or ornaments to the singer's melody. Singing and moving together as one may be one of the deepest ways political order is consummated; Plato himself brings his ideal city into life through singing and dancing. Thus, that city would best be nurtured by all its citizens singing a melody in unison, perhaps underlined by instruments that do not seek to undermine the unanimity of that song by adding a variant or different voice.[27] Thinking back to Plato's concept of the three parts of the individual soul, *heterophonia* might encourage conflicts between those parts within each soul as well as in the city.[28]

The evidence of ethnomusicology indicates the impossibility of distinguishing "natural" from socially ordained behavior. It may be as "unnatural"—or as "natural"—to make a group of singers achieve a completely unified unison as it is to make them sing divergent melodic lines. Achieving either extreme requires discipline and practice. The consummate rendition of monophonic Gregorian chant with seamless ensemble and unanimity requires expert coordination, skill, subtlety, and taste. The difficulties of polyphony are more evident; a group of people who could sing passably in unison might well struggle to sing separate, independent parts. The training of truly independent voices requires each to maintain its own line while coordinating it with others.

In the case of ancient Greek musical practice, our knowledge has grown in the past century, thanks to rediscovered manuscripts. The surviving ancient musical texts use a special code of alphabet-like symbols written above the sung syllables, lacking the customary visual clues of "rising" or "falling" melodic line that singers can use to grasp the "shape" of the melodic line intuitively, as in modern staff notation.[29] For instance, consider what may be the only surviving complete piece of Greek music, a short drinking song (*skolion*) found on the gravestone of one Seikilos (about 100 CE; figure 1.4; ♪ sound example 1.3), a call to *carpe diem* meant to console the passerby as well as those whose grief commissioned this carving, though the stone survived the passing of many days since then.

This alphabetic-symbolic notation called for a trained "reader" to decode it, indicating that musical literacy was reserved for the few, perhaps to guard the guild secrets of the bards. Very few such readers (if any) could have sight-read an unknown melody, as Guido of Arezzo, the eleventh-century inventor of staff notation, boasted his pupils could; the ancient and modern notations of the Seikilos song in figure 1.4 show a difference in transparency comparable to Roman numerals next to Hindu-Arabic numerals.[30] Probably the ancient Greek notation was meant only to remind those who already knew the tune, rather than teach it to those who did not.

The ancient system of notation was well suited to monophony because each syllable of text is naturally matched with a single sequence of melodic symbols written above it. Any other simultaneously sounding pitch would seemingly require another horizontal register beside the two already present. That no other extra registers have been found both confirms the basic monophony of ancient Greek music and also reinforces the insight that it was deeply dependent on its verbal text: the alphabetic musical notation acts as a kind of metatext written above the text whose melody it notates, as if indicating the close union of both within a single melody (*melos*).

Figure 1.4
The Seikilos *skolion* (a) as it appeared on the front and back sides of a cylindrical marble stele, here shown next to each other; (b) transcribed in its original notation; (c) transcribed into modern notation (♪ sound example 1.3). Text: "While you live, shine. Don't suffer at all; life is short and in the end time will have its way."

Even apart from this melodic notation, spoken Greek syllables have built-in rhythms, inherently short and long in quantity, as well as intrinsic pitch-accents, which could be independent of the musical line. For instance, the spoken word *zês* ("you live") normally had a rising, then falling pitch indicated by a circumflex over the *ê*; in the Seikilos song, this word is set to a single pitch. Thus, the Greek language has an inherent "melodic line" that is so musically clear that the overlay of additional melodies might well have been felt as excessive or detractive from the melodic qualities of the single line of text alone.

In light of this, scholars were surprised to find in a fragment of Euripides's *Orestes* certain notation-symbols without any text syllable underneath; the current consensus seems to be that these may represent notes of the accompaniment, perhaps giving cues for the aulos players who were known to have accompanied such tragic choruses (figures 1.5–1.6; ♪ sound example 1.4).[31] Current scholarship interprets these extra notes as drone pitches, sustained by the instruments against the moving choral line. Nor does this secondary pitch remain fixed; for several measures, the drone changes as the melody itself shifts, indicating a certain dynamic function for the changing drone-notes, which here act to underline a melodic shift as well as shifts in the mood of the text.

This might have been acceptable to Plato's Stranger, for in every visible respect these additional notes (whether sustained as drones or not) seem to frame the main musical line and its text, which remains the predominant musical determinant throughout. Perhaps the treatises considered such auxiliary notes to be so universally known and presumed that they

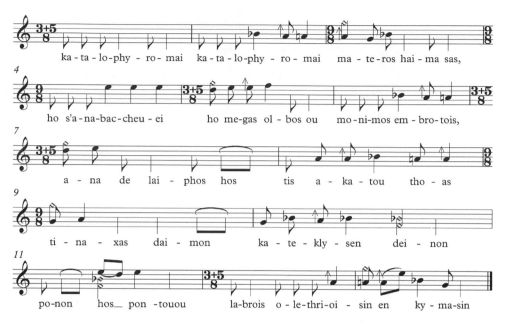

Figure 1.5

Euripides, *Orestes*, lines 338–344 (♪ sound example 1.4), a modern transcription of a fragmentary score copied about 200 BCE, indicating by headless notes words whose musical setting was missing on the papyrus, along with the missing syllabic lengths. The symbol ↑ denotes pitches raised by a quarter tone (diesis). Diamond-shaped notes were played by an aulos accompaniment, according to Pöhlmann and West (2001). Text (Chorus of women of Argos): "I grieve, I grieve—your mother's blood that drives you wild. Great prosperity among mortals is not lasting: upsetting it like the sail of a swift sloop some higher power swamps it in the rough doom-waves of fearful toils, as of the sea."

did not merit comment because they were not substantively independent of the melody. At one point in his *Manual of Harmonics*, the Greek theorist Nicomachus (writing in the first century CE) explicitly referred to striking "two strings simultaneously," though most of his other references seem to concern successive, rather than simultaneous, sounds.[32] Indeed, the auxiliary notes in the *Orestes* chorus really do not form a melodic voice of their own, but remain accompaniment underlining the vocal line. Though the Stranger's critique points to the existence of more daring *heterophonia* than is evidenced in the few surviving manuscripts, his description does not go further than indicating more or less ornate instrumental accompaniment to a vocal line.

Plato held that both the soul and the cosmos were made out of music. Against the prevalent monophony of Greek music, Plato's description of the "music of the spheres" may be the first place where polyphony comes forward not just as a peripheral or incidental musical practice but as an essential attribute of the cosmic harmonies.[33] Plato's *Republic* ends with a mythic description of a voyager named Er looking down on the heavens, viewing the cosmic spindle of Necessity and hearing its circular whorls, each associated with a

Figure 1.6
An aulos player accompanying a dancer. Associated with Dionysus, the aulos was a reed instrument with a nasal tone, often played in pairs.

planet: "And up above on each of the rims of the circles stood a Siren, who accompanied its revolution, uttering a single sound, one single note. And the concord [*symphonein*] of the eight notes produced a single harmony [*mian harmonian*]."[34] Up to that point, *harmonia* (literally a joint, as between ship's planks, or a framework, an agreement) in Greek seems generally to have meant a single melodic line, as opposed to our word "harmony" with its predominant sense of blending different notes into a simultaneous chord.[35] Accordingly, Plato's description has often been interpreted as describing some kind of melodic line drawn from the several notes of the Sirens.

Nevertheless, his text emphasizes that each Siren (or planet) uttered "a single note" and that all eight together produced "a single *harmonia*" as "the concord of the eight notes." The evident deduction is that the planets are producing an eightfold polyphony, at least in the sense of eight simultaneously sounding notes, a chord (as we would call it), not just a melody composed of those notes in succession. Plato's specific and pointed language also suggests that he is describing something previously unknown or at least undescribed. If so, in this text "harmony" in the sense of polyphony may find its earliest expression as an important new element of music and astronomy as sister sciences. Having banished polyphony from earthly song, Plato reserved it for the cosmic harmony.

Plato's description also makes clear that this celestial music is utterly different from the ancient music of the passions whose powers he sought to regulate and restrain. His voyager looks down on "a straight column of light that stretched over the whole of heaven and earth, more like a rainbow than anything else, but brighter and more pure."[36] The music he hears is made by Sirens, not humans; he beholds "the Fates, the daughters of Necessity: Lachesis, Clotho, and Atropos. They were dressed in white, with garlands on their heads, and they sang to the music of the Sirens" (figure 1.7).[37] We stand far above the Olympian gods, farther still above humanity; there is no hint of passion or emotion, only the sublimity and awe of the cosmic harmonies to which the Fates make their song, in which "Lachesis sang of the past, Clotho of the present, and Atropos of the future."[38]

Roman commentators also drew attention to this strange new kind of music. In his *Republic* (written about 54–51 BCE), Cicero included a section on "The Dream of Scipio" (*Somnium Scipionis*) that recounted Plato's story, including the polyphony of the planets: "It is that which is called the music of the spheres, being produced by their motion and impulse; and being formed by unequal intervals [*intervallis coniunctus inparibus*], but such as are divided according to the most just proportion, it produces, by duly tempering acute with grave sounds, various concerts [*concentus*] of harmony."

In the earthly realm, Cicero underlined the political implications of the harmony between different voices. Though Plato had only mentioned polyphony among the planets in his *Republic*, Cicero makes explicit the polyphonic aspects of the musical structure he considers important for the harmony of an ideal commonwealth:

For just as in the music of harps and flutes or in the voices of singers a certain harmony of the different tones must be preserved, the interruption or violation of which is intolerable to trained ears, and as this perfect agreement and harmony is produced by the proportionate blending of unlike tones, so also is a *civitas* made harmonious by agreement among dissimilar elements, brought about by a fair and reasonable blending together of the upper, middle, and lower classes, just as if they were musical tones. What the musicians call harmony in song is concord in a *civitas*, the strongest and best bond of permanent union in any commonwealth; and such concord can never be brought about without the aid of justices.[39]

Perhaps because he is concerned with a much larger commonwealth (*civitas*) than the ideal city (*polis*) Plato had described, Cicero brings forward the diversity of the voices both in the music of his time and in the political harmony to which he analogizes it.

Following in the Platonic tradition, Martianus Capella emphasized the view of celestial music as "harmony" in our sense of simultaneous sonority in a later Roman work, *The*

Figure 1.7
Bernardo Buontalenti, *Anankê and the Fates* (1589), a drawing for "The Harmony of the Spheres," the first *intermedio* performed at the wedding of Ferdinando de' Medici and Christine of Lorrain (see also figure 8.6). Note the open mouths of the three Fates, who are singing.

Marriage of Philology and Mercury (written sometime between 410 and 439 CE, not long after the sack of Rome by Alaric in 401). This densely allegorical and seemingly pagan work became a touchstone for many writers in the following millennium. In it, Mercury, the messenger god (representing intelligence), seeks a bride but is successively rejected by Wisdom, Divination, and the Soul. To find divine advice, in a mysterious cave Mercury meets Apollo, who calls up visions of the physical world, both past and present, including a musical grove in which

a tuneful melody caused by the whispering winds in the trees rustled with a certain musical vibration. The topmost layers of the tall trees, correspondingly stretched tight, reverberated a high sound, but whatever was close and near to the ground resounded a deep, heavy note through the down-turning branches. The middle portions of the trees, coming in contact with each other, sang together in the accompaniments [*succentibus*] of the octave [whose ratio is 2:1], the fifth [3:2], the fourth [4:3], and even the whole tone [9:8], without any discontinuities, as long as the semitones were included. In this way, the grove sounded the full harmony and song of the gods in melodic concordance. When the Cyllenian [Mercury] explained this, Virtue became aware that even in the heavens the spheres produce harmony according to the same ratios or combine with other voices in accompaniment [*succentibus*]; so it is not strange that the grove of Apollo should be so full of harmony, when the same god, in the sun, modulates the spheres of the heavens also.[40]

Thus, we learn that some planetary spheres *accompany* others in the celestial harmony, whose simultaneous sound is specifically confirmed in the striking image of the grove of trees reverberating that heavenly music in low, middle, and high registers *at the same time*.

Martianus makes even clearer his awareness of polyphony in his description of the feast at which Harmonia appears as the last of the prospective brides for Mercury: "Immediately a sweet new sound burst forth, like the strains of auloi [see figure 1.6], and echoing melodies, surpassing the delight of all sounds, filled the ears of the enchanted gods. For the sound was not a simple one, monotonously produced from one instrument, but a blending of all instrumental sounds creating a full symphony of delectable music."[41] Martianus emphasizes the newness of this kind of simultaneous music-making, whose novelty and extravagance seems to make it fit for the gods. When Harmonia herself appears, she carries a circular shield

with many inner circles. ... The encompassing circles of this shield were attuned to each other, and from the circular chords there poured forth a concord of all the modes. ... All other music—which, by contrast with its sweetness, seemed dissonant—now became silent. Then Jupiter and the other heavenly beings, recognizing the grandeur of the more exalted melodies, which were pouring forth in honor of a certain secret fire and inextinguishable flame, reverenced the profound ancestral song, and one by one arose in homage to extramundane intelligence.[42]

Martianus's description is suffused with his Neoplatonism, the tradition that interpreted Plato to have taught that the divine One overflows to produce the Many, including all the various things in the world. Martianus shows the gods paying homage to the primal Mind, one of the highest outflowings of the One according to Neoplatonists such as Plotinus.[43] Martianus gave no evidence of being a Christian; like Plotinus, he may even have opposed the new religion. In accord with his apparent polytheism, Martianus asserts that this "extramundane intelligence" is fundamentally polyphonic, underlying the whole cosmos and the gods themselves. Harmonia explains that she is to be the governess of the souls emanating from the primal source, "the One," whose overflowing brings them to life.[44] Her shield pouring forth "the profound ancestral song" symbolizes the Neoplatonic understanding that the different polyphonic voices flow from the One as their source and origin. The listening gods recognize that the monophony they knew, however unified, seems dissonant compared to the sweetness of polyphony, which they acclaim as if it were both the newest and oldest music, worthy "of a certain fire and inextinguishable flame."

Thus, though Plato defended purely monophonic music as most appropriate to human affairs, he associated the divine music of the spheres with a new kind of polyphony, whose many simultaneous pitches reflected the cosmic harmony. Cicero connected this celestial polyphony with the harmonization needed by the larger commonwealth of Rome and its burgeoning empire, so much vaster than the compact city-states of Greece. After the fall of Rome, Martianus Capella presented this Platonic polyphony, so different from the music generally known, as appropriate to the "extramundane intelligence," the highest divinity.

2 Dispassion and Deification

The music of early Christian worship was monophonic and sought to lay the passions to rest, setting itself apart from pagan displays of emotion with their theatrical associations. Thus, Gregorian chant did not indulge in expressive word painting or overt emotionality. By the ninth century, polyphony entered liturgical use; its relation to prayer and divinity came under increasing scrutiny. Its justification touched on sensitive theological issues that were the subject of continuing and fierce contention. These questions circled around the reformulation and reconsideration of the general concept of *person*, whose status was at stake in these theological debates. Over several chapters we will follow the polyphonic "person" comprising many individual voices. Looking back to Plato and Martianus, the medieval followers of Neoplatonism were particularly sympathetic to polyphony. At the same time, important Christian apologists took up the language of polyphony to express difficult theological concepts. The underlying tensions between their arguments laid the basis for later confrontations.

Surrounded by pagan music, early Christian worship chose to avoid its sensualism and emotion. In the fourth century, Niceta of Remesiana stipulated that chant should

be sung in a manner befitting holy religion: let it not display theatrical turgidity, but show a Christian simplicity in its melody, and let it not evoke the stage, but create compunction in the listeners. Our voices ought not to be dissonant but concordant—not with one dragging out the song, and another cutting it short, while one sings too softly, and another too loudly—and all must seek to blend their voices within the sound of a harmonious chorus, not to project it outward in vulgar display like a cithara. It must all be done as if in the sight of God, not man, and not to please oneself.[1]

His contemporary John Chrysostom (a famous preacher) emphasized the importance of monophony for the Christians: rising before dawn: "radiant and cheerful, they form one choir, and all together in unison [*symphōnōs*] and with a clear conscience, they sing, as if from one mouth, hymns to the God of all," like "a choir of angels."[2]

During that same period, Evagrius of Pontus left a worldly life in Constantinople (where he was involved in an affair with a married woman) to become an ascetic in the Egyptian desert. In his new life, he put much importance on the chanting of psalms, advising his brethren to "pray with moderation and calm, and chant psalms with understanding and proper measure [*eurhythmōs*], and you will be raised on high like a young eagle. Psalmody lays the passions to rest and causes the stirrings of the body to be stilled; prayer prepares the mind to perform its proper activity."[3] Evagrius emphasized that prayer and music should rely on calm, unimpassioned moderation, rather than emotional response. His asceticism might best be understood by a comparison he makes with athleticism: the athlete and the ascetic both avoid passion and sensual indulgence not because they are bad but to intensify their physical or spiritual training. Evagrius avoided passion because he judged that it "stirs up" the body through the emotions in such a way that the soul shifts its focus from God alone. In this, he followed the long-standing philosophic advice of the Platonic tradition to contemplate the Good with the eyes of the intellect unclouded by passion.

Evagrius's counsel to avoid passion presumes earlier stages of spiritual development that acknowledged and confronted the passions. Those who are just beginning their life of prayer should first of all address their impulses to gluttony and the other "eight thoughts" or vices, especially "sadness," the dejection of the soul resulting from anger and the frustration of pleasures. "When the soul has been purified by the full complement of the virtues [commandments], it stabilizes the attitude of the mind and prepares it to receive the desired state," which Evagrius compares to Moses approaching the presence of God: "If Moses, when he tried to approach the earthly burning bush, was held back until he removed the sandals from his feet, how can you, who wish to see and commune with the one who is beyond all representation and sense-perception, not free yourself from every mental representation tied to the passions?"[4]

Even after penitents shed their gross faults, they may fall into more subtle spiritual temptations, particularly pride accompanied by self-righteous anger. To purge these temptations, Evagrius advises that one should "pray first to receive tears, so that through compunction you may be able to mollify the wildness that is in your soul." Lest this itself become a further point of pride, "even if you pour forth fountains of tears in your prayers, entertain absolutely no exaltation within yourself for being superior to most people."[5] Evagrius often uses the word *catharsis* for purification through tears and self-discipline, using Aristotle's word describing the positive effect of tragic drama. Though often translated as "purgation," with the medical implication of expelling some disturbing or pathological matter, this Greek word also connotes winnowing or sifting, cleansing, or even explaining. Catharsis restores the soul to its original pristine state by removing the clouding caused by passion.

To the degree that disturbing passions are stilled, the soul can then enter into its proper activity, "the ascent of the mind towards God." Evagrius describes the realm of this "true prayer" as "an impassible habit [*apatheia*]" that communes with God "as with a father, while turning away from any mental representation tied to the passions."[6] He excludes all mental representations, not only the passions, because "contemplations of objects leave their impress and form on the mind and lead it far away from God," even if those objects are "simple intellections" of holy objects. For Evagrius, the state beyond passion (*apatheia*) is not dull apathy but a state of spiritual clarity he calls "supreme love."

In a similar spirit, Gregorian chant eschewed theatricality in favor of serene melody, reflecting the legend that the Holy Spirit in the form of a dove—not human artifice—imparted it to St. Gregory the Great (figure 2.1). Rather than expressively echoing the meaning of its text, a number of Gregorian chants reuse music to set notably different texts, sometimes with contradictory imagery.[7] For instance, the same Gregorian melody with a prominent rising scale was used to set the texts "I ascend to the Father" and "I went down into my garden."[8] Box 2.1 presents two examples of Gregorian chant and their melodic construction, which has more to do with its mode than its text.

Box 2.1
Gregorian chant

Consider the first verse of the chant "Kyrie cunctipotens genitor" (♪ sound example 2.1). In this Gregorian notation, the clef indicates the position of the note C:

Text: "Lord, have mercy." This chant is in mode one (called "Dorian" through a misunderstanding of the ancient Greek modal system, which would have labeled this "Phrygian"). In mode one, the chant is built around the notes D (the "final") and A (the "dominant" or "reciting tone"), a fifth above it; note the prominent minor third D–F. In this chant, after initially circling around A, the melody descends in small waves down to the final D.

Compare the following chant for Christmas Day, "Viderunt omnes" (♪ sound example 2.2):

Box 2.1 (continued)

Text: "All the ends of the earth have seen the salvation of our God: rejoice in God, all the earth." This second example is in mode five (Gregorian Lydian mode), whose final is F and dominant is C. This chant emphasizes F–A–C, the gesture with which it begins, then spends the middle section ("omnes … Dei nostri") moving in various ways around the dominant (C); by comparison with the minor third in "Kyrie cunctipotens genitor" note here the major third F–A. The final words ("omnis terra") form successive waves that begin and end on the final (F).

Only much later did some writers interpret chant in terms of the passions, compared to this dispassionate tradition. For instance, around 1100 a certain John's *De musica* (*On Music*) entertained the possibility of composing new chants, "even if new compositions are now not needed for the Church." Such chants should "be varied according to the meaning of the words. … If you intend to compose a song at the request of young people, let it be youthful and playful, but if of old folk, let it be slow and staid."[9] He thought inauspicious texts should be pitched low, auspicious ones high; John cites examples of Gregorian chants (with texts on David's grief and the resurrection, respectively) to illustrate his point.

If, as ethnomusicologists argue, various forms of polyphony are widely practiced, then we should not be surprised that the Western European tradition also included polyphonic developments.[10] Describing informal music-making in the British isles in 1198, the Welsh churchman and historian Gerald de Barri noted that

when they made music together, they sing their tunes not in unison, as is done elsewhere, but in part with many simultaneous modes and phrases. Therefore, in a group of singers (which one very often meets with in Wales) you will hear as many melodies as there are people, and a distinct variety of parts; yet, they all accord in one consonant and properly constituted composition. In the northern districts of Britain, beyond the Humber and round about York, the inhabitants use a similar kind of singing in harmony, but in only two different parts, one singing quietly in a low register, and the other soothing and charming the ear above. This specialty of the race is no product of trained musicians, but was acquired through long-standing popular practices.[11]

Gerald also noted their "rapid and lively" playing of musical instruments, suggesting some interaction between vocal practices and the polyphony of instrumental ensembles. Monastic novices might have been familiar with informal part-singing in their villages, which choirmasters may have discouraged as they trained them to conform to the prescribed practice of singing in disciplined unison with the other choir monks, to sing the monophonic chant that was the prevalent ecclesiastical practice.[12] Indeed, Charlemagne himself ordered the standardization of the Roman chant throughout his domains. Yet the temptation to try a bit of polyphony might have remained, or perhaps was an alternative tradition that never really died, given the general tendencies toward heterophony that ethnomusicologists have noted.[13]

Figure 2.1
A dove imparts a chant to St. Gregory the Great, who dictates it to a scribe, from the Antiphonary of Hartker in the monastery of Saint Gall (ca. 1000). The historical Gregory, however, lived about two centuries after the earliest recorded Gregorian chants.

Thus, early ecclesiastical polyphony possibly reflected a new synthesis with secular music, of which little written evidence survives. In general, though, the extant texts seem to privilege influence flowing from sacred to secular, not vice versa. Our question is not so much "Why polyphony?" if indeed it was part of general tendencies in that direction. Instead, our question is why was polyphonic music not rejected as tainted by popular practice but rather advocated by some learned clerics, who mandated it for great ecclesiastical feasts?

In the Western Catholic liturgies, beginning at least in the ninth century polyphony was considered to add "solemnity" in a technical sense, an elevation of protocol appropriate to certain important ecclesiastical services. Liturgical solemnity specifically calls for several independent personages (such as the celebrant, deacon, subdeacon) to take different parts in the service, as well as the choir that would participate in an ordinary high (sung) mass; the issue of choral versus solo singing was important for the history of psalmody from its very beginning.[14] From early Christian times, the choir itself was often divided into two subchoirs that would alternate responsively, answering back and forth across the church, thus providing another possible invitation to multiplicity of vocal parts. Thus, ecclesiastical solemnity has a kind of incipient or potential polyphony in its multiple clerical voices, which might have encouraged the use of polyphonic music for such services.

A common Neoplatonic heritage may help explain the apparent coincidence that, at nearly the same time as Martianus or perhaps a few years earlier, musical simultaneity emerged at the center of a seminal Christian text. Augustine of Hippo's discussion of the nature of the divine incarnation in his *De Trinitate* (*On the Trinity*, written in the first decades after 400 CE) draws on his own extensive musical and rhetorical education as a brilliant young pagan in Carthage. About ten years earlier, prior to his conversion to Christianity, he had written a short treatise *De musica* (*On Music*), whose surviving text concentrates on poetic rhythm and prosody, rather than melody, and which testified to the depth of Augustine's musical studies and interests.[15] In his later description of the Trinity, he called on these concepts at the crux of his discussion of how the divine Word could become flesh; among his theological works, this is "one of the very rare allusions to his musical knowledge," as his modern editors note.[16]

Ancient Greek music theory ascribed to Pythagoras the discovery that two strings at equal tension sounded an octave when their lengths were in the ratio of 1:2. Using this celebrated idea, Augustine used the accord between "single" and "double" in this octave to describe "how the single [*simplum*] of our Lord Jesus Christ matches our double [*duplo nostro congruat*], and in some fashion enters into a harmony of salvation with it."[17] That is, the relation of Christ to ordinary humans "matches" the octave, 1:2, as if his saving grace sounded the octave above our humanity. Though Augustine was steeped in Neoplatonic ideas of the One and of the Dyad (the first mysterious outflowing from pure unity toward multiplicity), there is no precedent in earlier Neoplatonic writings for his daring metaphor of the octave, which incarnates numerical concepts in audible sounds to express analogically how Christ is the Word made flesh.[18] In Augustine's view, though the Neoplatonists seemed to have found their own way to some of the deepest Christian teachings, they did not recognize the incarnation of Jesus because their philosophy could not accept that the mysterious emanations of the divine One could ever have become human flesh, as if the divine and the human were only separated by an octave.[19]

Returning to his description in *De Trinitate*, Augustine emphasized his own struggle to find the right word to describe the union of Word and flesh. His deliberately groping rhetoric mirrors and intensifies the felt effect of his musical imagery, leading to the climactic point at which he adapts a new term to match his sense:

This match—or agreement or concord or consonance or whatever the right word is for the proportion of one to two—is of enormous importance in every construction or interlock [*coaptione*]—that is the word I want—of creation. What I mean by this interlock, it has just occurred to me, is what the Greeks call *harmonia*. This is not the place to show the far-reaching importance of the consonant proportion of the single to the double. It is found extensively in us, and is so naturally ingrained in us (and who by, if not by him who created us?), that even the unskilled feel it whether singing themselves or listening to others. It is what makes concord [*consonantia*] between high-pitched and deep voices, and if anyone strays discordantly away from it [*dissonuerit*], it is not our knowledge, which many lack, but our very sense of hearing that is painfully offended. To explain it would require a long lecture; but anyone who knows how can demonstrate it to our ears with a monochord.[20]

To illustrate this "interlock," Augustine gives the example of high- and low-pitched voices (such as boys and men) singing in octaves. Though the singers may not be educated and do not know the mathematics Augustine mentions, they feel acutely and immediately whether their octave is exactly in tune. Indeed, such practices of singing at the octave are almost inevitable (and often happen unintentionally) when such mixed groups sing together whose vocal ranges are sufficiently diverse (such as boys and men or men and women). From more learned practice, he specifically mentions the monochord (figure 2.2), whose measured and weighted strings were used to illustrate musical ratios beginning with the ancient Pythagoreans. Thus, Augustine showed awareness of the monochord's existence and use a full century before Boethius, who is usually credited as the first to mention it in the West.

Elsewhere, Augustine uses the specific sense of harmony as accord between different voices to express the ideal political community enjoyed by righteous souls. On the Day of Judgment, he envisions that the accord that always ruled the heavens will then also rule on Earth: "Even then the just of God [*sancti Dei*] will have their differences, accordant, not discordant, that is, agreeing, not disagreeing, just as sweetest harmony arises from sounds differing indeed, but not opposing one another."[21] Their concord comes from the simultaneous blending of their differences, not from absolute unanimity or unison.

Likewise, Augustine's use of the image of octave singing makes clear that Christ, as the One, "matches our double" *simultaneously*, for if not, his divine nature would never finally

Figure 2.2
Pythagoras demonstrating musical consonances on monochords, from Franchinus Gaffurius, *Theoriae musicae* (1492).

blend with ours and we could never "participate in the Word, that is, in that *life which is the light of men*," as he puts it. Thus, Augustine used our natural reaction to the octave to show that we have an innate awareness that is not externally learned; his daring comparison of the octave to the "interlock" (*coaptione*) between soul and Christ expresses the divine origin of both, for without such a divine source inspiring us, how else could our flawed human understanding reach such acute awareness?[22] He takes the simultaneous octave to symbolize Christian incarnation and human redemption as united in *harmonia*. Compared to the Neoplatonists, Augustine is far more interested in the physicality of sound precisely because he understands it as the perfect image of the union of human and divine natures essential to Christian teaching. At the same time, he retains the Neoplatonic argument that "all that is eternal is a simultaneous whole."[23] Augustine incorporates simultaneity in his musical metaphor, which portrays the human soul tasting eternity in simultaneous sonorities, reverberating the eternal Word.

Augustine here addressed what was arguably the most difficult and controversial theological question of his time, the relation between the divine and human natures in the person of the Christ. By the First Council of Nicea (325), the essential outlines of the doctrine of the Trinity had been agreed by the Eastern and Western churches, though Augustine himself was involved in further controversies about the procession of the Holy Spirit that to this day remain points of disagreement between those churches. Yet as difficult as was the concept of the Trinity, the problem of the two natures of Christ was even more vexed. During Augustine's lifetime and into the centuries beyond it, those who were called Arians contended that Christ was a created, though divine, being, who was subsidiary to the Father; Trinitarians held that the Son was not subordinate and had always existed with the Father. This controversy divided the Christian world far more urgently than other issues about the Trinity; some powerful ecclesiastical and political figures took up the Arian cause, including barbarian warlords and Roman emperors.

The passage in Augustine we have just been considering used the musical analogy of the octave to express his view of Christ harmonizing divine and human natures. In this context, Augustine carried his argument further by using the term "person," introduced by Tertullian as a way to characterize the differences between Father, Son, and Holy Spirit, lest they seem merely different aspects of a single entity, rather than three distinct beings. To avoid getting lost in the complexities of terminology and theology, I will summarize the salient points, giving further details in box 2.2.[24]

Box 2.2
Being and Substance

The Greek noun *ousia*, derived from the verb "to be" (*einai*), has the general sense of "being-ness." Originally it meant "that which is one's own, one's substance, property," as in the older English phrase "a man of substance," someone who has substantial assets, particularly land. From this, *ousia* came to denote stable being, immutable reality, rendered in Latin sometimes as *essentia*, sometimes as *substantia* (English "essence" or "substance"). On the other hand, *hypostasis* literally means "standing under [sub-stantia], supporting," such as the foundation on which a temple stands or the title deeds supporting a claim of ownership. Thus, the Greek distinction between *ousia* and *hypostasis* did not map smoothly into *essentia* and *substantia* or "essence" and "substance."

When Augustine was trying to explain "three what?" are in the Trinity, he found "obscure" the distinctions that the Greek language makes between "being" (*ousia*) and "substance" (*hypostasis*), concepts Latin conflated into one word (*substantia*). Instead of "one being or substance, three persons is what many Latin authors, whose authority carries weight, have said when treating of these matters, being able to find no more suitable way of expressing in words what they understood without words."[25] Thus, Augustine adopts the term "person" to conform to the usage of other Latin authors, noting that Greek authors do not use this term.[26] He is aware that "scripture calls these three neither one person nor three persons—we read of *the person of the Lord*, but not of the Lord called person—we are allowed to talk about three persons and the needs of discussion and argument require; not because scripture says it, but because it does not gainsay it. Whereas if we were to say three Gods scripture would gainsay us, saying *Hear, O Israel, the Lord your God is one God*."[27] In his view, the word "person" is not the only possible term but rather a concession to the inadequacy of human words in response to "the sheer necessity of saying something, when the fullest possible argument was called for against the traps or the errors of the heretics."

The concept of personhood resonates in many different aspects of Augustine's works. His famous *Confessions* are also notably personal, in the modern sense of that word, vivid and emotionally fraught, revealing many aspects of his inner and outer lives that he considered shameful but essential parts of his spiritual journey. In so doing, Augustine disclosed a new depth of interiority.[28] Yet, though he was arguably the first thinker in the West to emphasize and analyze the concept of person, Augustine left it sufficiently undefined that it came under the further scrutiny of those who followed him.[29]

According to a traditional etymology, the Latin word *persona* derives from the mask worn by an actor in tragic drama, the visible facade through which the voice sounded (*personare*).[30] *Persona* came to have the legal meaning of someone who could appear in the "theater" of the law courts, address the court, give evidence, take oaths, and so forth. In Roman times, personhood at first was a high and exclusive distinction, reserved in each family only to the *paterfamilias*, not to his wife, sons, daughters, or slaves unless he conferred it on them in a special ceremony conducted before an appropriate tribunal.[31] Nor was personhood limited to individual human beings; Roman law began the concept of corporation, an "artificial person" that could be represented in court and was immortal (unlike "natural persons"). Even a statue in a temple could become an artificial person, its interests protected and represented by counsel.[32] Against this legal background, the concept of person became an important point of reference as Christians struggled to distinguish their novel concept of the godhead from polytheism and to clarify the distinctions within the Trinity. Personhood also provided a possible way to understand the interrelation of the two natures of Christ, divine and human.

About a century after Augustine, these issues remained hotly contested. As he addressed them, Anicius Manlius Boethius presented himself as a translator rather than an innovator; writing after the fall of Rome at the hands of barbarian invaders and as the chief minister to the Ostrogothic king Theodoric, he felt the need to record Greek philosophy in the Latin language, fearing that it would soon be lost. He was able to complete only a small part of his plan. As a Christian, Boethius also tried to define personhood more precisely in his theological writings. In addition, his *Fundamentals of Music* became the most important text transmitting ancient musical theory for the succeeding millennium. Thus, to his later readers, his works would have transmitted both the concepts of music and of personhood. Boethius was trying to mediate between Theodoric, who was an Arian, and the orthodox emperor, seeking a compromise for the controversy about the two natures of Christ. Theodoric took these activities as evidence that Boethius was conspiring against him with the emperor and condemned him to death; written in prison, Boethius's famous *Consolation of Philosophy* connects the pathos of his personal situation with the philosophic tradition from which he drew sustenance.

Though he attributes his inspiration to Augustine's *De Trinitate*, in his dialogue *Contra Eutychen et Nestorium (Against Eutyches and Nestorius)* Boethius specifically grappled with the problem of the dual nature of Christ as human and divine, for which he provided a definition of person as "the individual substance of a rational nature." Even as he followed Augustine, Boethius's definition stressed both individuality and rationality because he judged that a stone, a tree, or a horse cannot be a person, "but we say there is a person of a man, of God, of an angel."[33] Using this criterion, Boethius then rejected the Nestorian heresy that the two natures in Christ imply that he is in fact two persons. In that case, Boethius objected, there would be no union between those two persons, human and divine, only a mere juxtaposition; hence "Christ is, according to Nestorius, in no respect one, and therefore He is absolutely nothing." We can only conceive that Christ saved the human race, Boethius insisted, if he indeed united human and divine natures in his one person. On the other hand, the single personhood of the Redeemer does not imply that he had (as Eutyches had argued) only one single nature that utterly absorbed his human nature into his divine nature.

For Boethius, personhood was precisely the way in which a duality of natures can still subsist in the singular person of the Christ, whose oneness Boethius considered a precondition for human salvation and the exemplary model for human identity in general. Though Augustine reserved the term "person" for the exalted individuals in the Trinity, Boethius also used it for angels and human beings, thus extending the concept of personhood widely. He also emphasized that *personare* comes from *sonus*, "sound," "and for this reason, that the hollow mask necessarily produces a larger sound," so that these amplifying masks help us recognize the voice of each personage in a tragedy. Thus, Boethius explicitly directs us to *sound*, and hence implicitly to *voice*, as essential to personhood.[34] This connection will remain important as we consider the application of these concepts to polyphony.

Four centuries passed before the ideas we have been discussing came to wider distribution; only in the ninth century did Boethius and Martianus Capella become available again in manuscript, at about the same time as the earliest written examples of polyphony called *organum* were presented in the *Enchiriadis* texts: *Musica enchiriadis (Handbook of Music)* and *Scholica enchiriadis (Commentary on the Handbook)*.[35] These manuscripts described practices probably already well-established as oral traditions before they were written down. Because the earliest surviving written evidence of Western chant is also roughly contemporary with these treatises, we also cannot assume that monophony simply preceded polyphony; both traditions coexisted, though ecclesiastical practice gave a special place to the monophonic chant.[36] Then too, these anonymous treatises seem to have been written somewhere to the north of the Carolingian Empire, perhaps codifying contemporary practices for the benefit of these more remote regions, rather than themselves making any claim to innovation.[37]

In these works, organum meant simultaneously singing a monophonic chant (the "principal voice") along with other voices (called "organal") to form "symphonies" (*symphoniae*), a term we will consider shortly. (The organ, as an instrument, was at that time referred to in the plural, *organa*; no extant evidence clarifies what relation, if any, it may have had to sung organum.)[38] Organum, in these early treatises, could include holding a drone note throughout the chant (a practice still observed in Greek Orthodox "*ison* chanting"); singing in parallel intervals (fifths or fourths) along with the chant; and freer motion in which the organal voice would sometimes parallel the chant, sometimes move obliquely against it (figure 2.3a,b; ♪ sound examples 2.3a,b).

The *Enchiriadis* texts came near the beginning of an extraordinary flowering of polyphonic music that involved the most innovative intellectual centers of Europe. We return to our question: How did the partisans of polyphony understand polyphony in relation to the monophonic tradition of chant? How did they conceive polyphony's relation to the sources and traditions they knew or to new intellectual and musical horizons? The *Enchiriadis* treatises most likely drew their connections between polyphony and theology from their main authority, Boethius, whom they call *doctor magnificus*. Where Boethius used the term

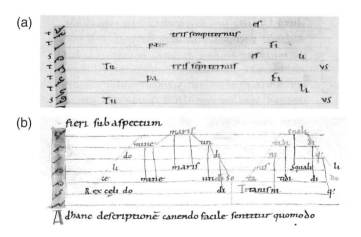

Figure 2.3
Organum from the *Musica enchiriadis*: (a) parallel organum at the fourth (♪ sound example 2.3a). Text: "You are the eternal son of the Father"; (b) free organum (♪ sound example 2.3b), in which the voices may move obliquely, rather than in strict parallel. The staff notation uses Dasein signs, an older form of notation than the present staff notation. Each Dasein sign denotes a pitch level in the mode. Text: "King of heaven, lord of the sounding sea, of the shining Titan sun, and the gloomy Earth."

symphonia to mean "consonance," the *Enchiriadis* treatises used this term to refer to "a sweet combination of different pitches [*vocum disparium*] joined [*concentus*] to one another," specifically when sounded simultaneously in organum. Though most of Boethius can be read as referring only to a single melodic line whose notes are sounded successively, not simultaneously, the *Enchiriadis* treatise quoted Boethius's explicit mention of two strings plucked at once, which he in turn drew from Nicomachus: when tuned an octave apart, two strings "combine and are united together in sound so that one pitch [*vox*], as if produced from one string and not mixed [*mixta*] from two, strike the hearing."[39]

"Thus says Boethius," *Musica enchiriadis* notes, invoking his authority on a matter it considers especially important and difficult: "why some tones agree with each other in a sweet commingling [*commixtione consentiant*], whereas others disagree unpleasantly, being unwilling to blend [*misceri*] with each other, has a rather profound and divine explanation, and in some respects is among the most hidden things of nature."[40] Likewise, Aristotle used "mixing" (*mixis*) to describe how "a concord is formed of high and low" sounds.[41] Note this language of "mixing" (*mixtere*); the agreement or disagreement of specifically *simultaneous* tones has a special importance for the *Enchiriadis* author, which he connected to larger issues: "This principle, whose operations in this realm the Lord also permits us to penetrate, is treated in the writings of the ancients. In these is asserted, with most convincing arguments, that the same guiding principle that controls the concord of pitches regulates the natures of mortals. Through these numerical relationships, by which unlike sounds concord [*concinentias*] with each other, the eternal harmony of life and of the conflicting elements of the whole world is united as one with material things."[42]

Hence, the *Enchiriadis* treatises connected the problem of concord and discord between simultaneous voices with "the natures of mortals," specifically with the way "the eternal harmony of life" is united with "material things." This exactly described the problem of the "concord" of the divine and human natures in the person of Christ; the text's reference here to the Lord (the first and only such mention in this treatise, except for its concluding prayer) also implied that these matters have to do with the Lord himself. The author of the *Enchiriadis* treatises was probably also aware of the theological works of Boethius (whose manuscripts were available even before those of his mathematical and musical writings).

Thus, these references to the problem of Christ's dual nature may well also refer to Boethius's formulation of personhood. If so, this contemporary evidence connected the problem of musical polyphony with the "theological polyphony" inherent in the person of Christ, giving us some indication of how this early treatise on polyphony thought about its larger theological and philosophic implications. In this nexus of interconnected references and problems, Boethius was a central figure. Was "concord" (of voices or natures) a *mixture* (in Greek *mixtis*) in which the combining elements retained their different identities, or a *blending* (*krasis*) in which those identities are lost and utterly fuse?[43]

Reading the *Enchiriadis* treatises in light of these distinctions invites us to view a polyphonic composition as a kind of "virtual person" because it unites several separate "natures" (here, the distinct voices) into one persona, the unified individuality of the musical work itself. Alternatively, each voice could be considered a separate "person" within a larger community; as our story advances, various authors will advance different readings. For Boethius, the encounter with Christ involves human and divine natures united within one person. For the *Enchiriadis* author, the encounter with polyphony "permits us to penetrate … that the same guiding principle that controls the concord of pitches regulates the natures of mortals."[44] In that way, polyphony became a privileged new way to apprehend the mysterious truths of theology.

Further, several contemporary authors directly connected polyphony with the text of Martianus, specifically his description of the mysterious grove of Apollo we considered in chapter 1.[45] In particular, the ninth-century Irish theologian Johannes Scotus Eriugena wrote a commentary on this scene describing the "accompaniment" (*succentus*) the planets Venus and Mercury made to "the sun, Saturn, Jupiter, and Mars with the firmament."[46] Thus, Eriugena connected Martianus's description to polyphonic practice.[47] Then too, Eriugena used the unusual term *coaptantur* to describe the "harmonizing" of "*organicum melos* composed of different qualities and quantities of voices," arguably referring to Augustine's deliberate use of this same term (*coaptatione*), noted above.[48] Though no direct evidence connects Eriugena with the *Enchiriadis* treatises, other arguments suggest he was aware of the contemporary musical practice of organum.[49] Indeed, Eriugena may have been one of Boethius's first readers, perhaps even the rediscoverer of his musical treatise.[50]

Above all, Eriugena can serve as the voice of the Neoplatonic tradition that welcomed polyphony because he asserted the fundamental unity and commonality of the divine and the human: "We should not understand God and the creature as two things removed from one another, but as one and the same thing. For the creature subsists in God, and God is created in the creature in a wonderful and ineffable way, making himself manifest, invisible making himself visible."[51] Following Maximus the Confessor, an important Greek theologian whom he translated, Eriugena argued that as God makes himself known to us through theophanies—manifestations of divine presence—we in turn can be exalted through *theosis* or deification, meaning "a conformation to God's Wisdom. What the intellect knows, it becomes. As Maximus says: air is not light; yet it is so filled with light that it seems to be nothing but light; iron molten in fire, although it remains iron, is indistinguishable from the fire. So will the creature be with God."[52]

This clearly implies that the human mind *becomes* the music it hears, an inextricable "blending" (*krasis*) in which individuality is ultimately lost in mystic union. Eriugena interprets the mediating activity of the incarnate Word to mean that "he, who from God made himself a human being, makes gods from human beings."[53] Therefore, Eriugena anticipates a return (*reditus*) "whereby the whole of creation, by an ineffable miracle, will be transformed into God," as Deirdre Carabine puts it—when human nature will return to angelic status, because (as Eriugena argued) paradise "signifies the human nature that was made in the image of God."[54]

To exemplify the ultimate unification of human and divine, Eriugena uses an explicitly musical metaphor:

Can we not apply the same principles to the human voice and the sounds of musical instruments: For every sound, whether of the human voice, or of the pipe, or of the lyre, retains severally its own quality while many of them in unity produce with suitable agreement a single harmony. Here also the argument from acoustics makes it clear that the sounds themselves are not confounded, although they are unified. For if any one of those sounds were to be muted, it alone will be silent, and none of the other sounds will supply the melody that came from the one that is now silent. From this it is clear that when it sounded with the others, it retained the property of its own quality. … From these and similar examples taken from the intelligibles and the sensibles you may easily see how there can be a unification of human nature without sacrifice of the properties of individual substances.[55]

If Eriugena learned from Martianus that the music of the divine mind is essentially polyphonic, then the human mind *becomes* polyphonic and divine by hearing organum, just as the listening gods recognized and acclaimed the sublime sounds in Apollo's sacred grove.

For Eriugena, hearing polyphony symbolized the divinization of humanity in virtually pantheistic oneness. Thus, the explosive theological issue underlying polyphony seems to have been the true relationship between the human and the divine, which orthodox teaching already considered sufficiently close that they could be unified in the person of Christ. Yet Eriugena went further still: everyone can be divinized, not just the Christ. Just as he argued that divine creation was a *necessary*—not simply voluntary—overflowing of the One, Eriugena also argued that infinite punishment and Hell were contrary to reason: instead, he held that "both good and evil will enjoy the spirituality and incorruptibility of body, the same glory of their nature, the same essence, the same eternity."[56] For the Western Church, this went much too far; at Councils in Vercelli (1050) and Rome (1059), Eriugena's teachings were condemned in the form in which they had been espoused by three of his followers.[57] Eriugena himself was condemned by a bull of Pope Honorius III in 1225.[58]

In my view, Eriugena's importance was not so much the precise details of his position (or to what extent it was or was not the exact source for the *Enchiriadis* treatises) but more that he was a prominent representative of the continuing influence of Christian Neoplatonism. He, along with a few others, kept alive the radical concept that the return from the Many to the One is essentially a "self-seeing," which Plotinus expressed as the ultimate vision of unity in the exalted closing of his *Enneads*: "In this seeing, we neither hold an object nor trace distinction; there is no two. The man is changed, no longer himself nor self-belonging; he is merged with the Supreme. … There were not two; beholder was one with beheld; it was not a vision compassed but a unity apprehended. … When the soul begins again to mount, it comes not to something alien but to its very self … the flight of the alone to the alone."[59] The Latin Church, on the other hand, came to regard the One as *not* our own "very self" in its supreme aloneness, beyond being, but as a wholly other Being—God—who ultimately meets us person to person, face to face.

These developments were not dispassionate or inevitable but historically and politically contingent; the 1225 papal condemnation of Eriugena partly reflected the wide circulation of his *Periphyseon* (*On the Division of Nature*) among the Albigensians (Cathars) in southern France, who were condemned as heretics in 1179 and attacked during the Albigensian Crusade (1209–29). Later, Eriugena's book was influential among the Beghards and other rebellious groups.[60] The political struggle of the central ecclesiastical and secular powers to subjugate these rebels therefore condemned their preferred theologian as well as the Neoplatonism and pantheism that could be invoked to justify the rebels' claims.

If Neoplatonism was the deepest current that approved polyphony, the opponents of the one would also likely oppose the other. In this controversy, the Latin concept of person was a useful theoretical construct that could preserve the unity of the Trinity and of the two natures of Christ while disallowing the most radical possibilities of human union with the divine, except in the unique Person of the Christ. In this way, the concept of person was turned from expressing the common state of human, divine, and angelic personhood instead to delineate and enforce the distinctions between them. As a musical corollary of these philosophical developments, the polyphonic mingling of voices also could become suspect.

3 The Music of the Blessed

During the twelfth and thirteenth centuries, polyphony swiftly took on ever more complex forms, especially through the composers associated with Notre Dame in Paris as its new cathedral was being built (figure 3.1). Though most ecclesiastical authorities at the time approved its use to add solemnity to great feasts, some prominent clerics and religious orders disapproved of the complexity of polyphony and preferred the simplicity of chant. Agreeing with Dominican teachings, Thomas Aquinas stated that human music ought to be simple (monophonic), though he argued that God and angels could know many things at the same time, as if their minds were inherently polyphonic. Despite his admiration for Thomas, Dante eschewed his musical rigorism and presented polyphony as appropriate not only for God and the angels but also for the blessed souls in paradise and, by implication, here on Earth. Dante made explicit the "personhood" of a polyphonic work by comparison to the theological union of divine and human natures in the person of the Redeemer, whom he called the "good Apollo."[1]

The *Enchiriadis* treatises, discussed in the previous chapter, recorded the state of organum in the ninth century; subsequently, polyphonic music flowered astonishingly. For instance, the note-against-note uniformity of the *Enchiriadis* examples gave way to freer, more complex possibilities. During the twelfth century, manuscripts associated with St. Martial in Limoges notate a style (later called "St. Martial organum") in which the long notes of a Gregorian chant in one voice are simultaneously accompanied by more florid motion in another, so that one note of chant may sound against many notes in the other voice (figure 3.2; ♪ sound example 3.1).[2] In this example, a single word of the chant text is stretched over many notes in the florid *discant*, as it came to be called, literally meaning "singing apart," which led to the expression "singing in parts." Yet it would be too simplistic to think of such more complex organum as "more developed" or "better" than older examples, as if these were successive stages on a track of progressive historical evolution, along which nineteenth-century thinkers tended to arrange successive musical or artistic styles, judging the earlier ones primarily as "more primitive" precursors of the later. Surely salient differences between styles were valued or used variously, depending on local tradition and ecclesiastical practice. Though many centuries removed, we can still try to recover

Figure 3.1
The Cathedral of Notre Dame in Paris, consecrated in 1182, in a fifteenth-century miniature by Jean Fouquet. The faithful are gazing at the hand of God as demons flee to the left and right. Text: "O God, come to my assistance; O Lord, make haste to help me."

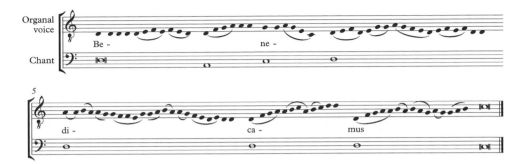

Figure 3.2
An example of St. Martial organum (♪ sound example 3.1). Text: "Benedicamus [domino]" ("Let us bless [the Lord]"). The long notes of the lower voice sound the Gregorian chant for this text.

in our own experience the various qualities evoked by these various kinds of polyphony: the more austere quality of the strict note-against-note style of the *Enchiriadis* treatises, compared to the more irregular fluidity of St. Martial style.

Such many-sided appreciation may help illuminate the richness of the further polyphonic innovations associated with what later scholars called the School of Notre Dame in the late twelfth century, which included the master composers named Leonin and Perotin.[3] For instance, in Leonin's setting of the Gregorian Gradual for Christmas Day, *Viderunt omnes*, the chant (discussed in box 2.1) is heard in long, unmeasured notes held in the lower of the two voices (figure 3.3; ♪ sound example 3.2). The chant-bearing voice became known as the *tenor* (from *tenere*, to hold). Above it, Leonin's organum weaves a free course that touches on the chant pitch but also uses many other notes freely against it, some quite dissonant (such as the initial E versus F dissonance in figure 3.3). Compared to the restrained pitch range of the chant, the organal voice's range is huge (an octave plus a fourth), its ornate line requiring a virtuoso executant, not just one of the anonymous choristers intoning the chant. To some extent, this development was already underway in the increasingly elaborate free organum of St. Martial. Yet Leonin's piece implies a whole new "political" reality of the soloist set against the chant (presumably intoned by a chorus), emphasizing the interplay between the rhapsodic elaboration of the organal voice and the sustained notes of the chant, contrasted against the choral voices singing the chant in unison. The sense of the chant syllable (much less the meaning of the complete word) is lost in the play of the organum. The timeless quality of the chant is further heightened by this fading of linguistic coherence: a single word of text now occupies so long a time that the mind only distantly tracks its meaning.[4]

This new polyphony required a momentous innovation, the development of a *rhythmic notation* that could coordinate the motion of these voices. It remains controversial whether Gregorian chant originally had some intrinsic rhythm; in recent centuries, the

Figure 3.3
Leonin, *Viderunt omnes,* based on the Gradual for Christmas Day (♪ sound example 3.2). Text: "All [the ends of the Earth] shall see [the salvation of our God: all the Earth shall rejoice in God]."

Roman Catholic Church tended to interpret chant notation as flowing in equal notes, rather than with some determinate rhythm. In any case, Leonin's notation specified the rhythms of his voices based on prosody, the long and short syllables in poetic meters, which Augustine had applied to music in his *De musica*.[5] But the greatest innovations came later and included notating the *rests*, the places where a voice would *not* sing, crucial to changing and coordinating rhythmic patterns. Perhaps here again Augustine gave inspiration with his charged references to silence as the condition in which the soul could transcend self.[6]

Rhythmic issues became even more important when Perotin set this same *Viderunt omnes* with four voices, rather than two, immensely extending each syllable of the chant; we hear forty measures of "*Vi-*" before changing to "*-de-*" (figure 3.4; ♪ sound example 3.3). Though using the same rhythmic mode or pattern as Leonin, Perotin's three organal voices interweave to form a very different texture, so closely spaced and overlapping in range that one often cannot tell one of them from the other. The ear is unsure exactly when the voices are intertwining and crossing or when they move apart. This confusion results partly from the greater multiplicity of voices: the ear attends differently to four voices than to two. Perotin called for a new texture of interwoven voices, as compared with the two distinct voices in Leonin. Here texture enters music in a new way, denoting complex effects emerging from the synthesis of a number of elements that together produce a net effect qualitatively different from their mere sum.

Compared to the soaring individuality of Leonin, Perotin's voices have a wholly other mode of being, independent yet indistinguishable in their intricate interconnection. Though their overall structural plans are similar, because the outer organum sections have been radically extended (to over a hundred measures, compared to Leonin's seventeen), one no longer perceives Perotin's structure in anything like the same way as one had Leonin's. Perotin's organum goes on at such length that one really loses oneself in it, far more so than in Leonin's much more compact setting. Some interpreters think Perotin intended his listeners to be overwhelmed and exalted, perhaps as one might feel when first entering a Gothic cathedral with much higher vaults and ceilings than one had ever before encountered.[7] Such a person might well feel disoriented, overwhelmed, perhaps experience a kind of vertigo gazing into the heights. Leonin and Perotin reduced their texts almost to incomprehensibility, at least to a human mind that can only understand words pronounced near their normal speeds, not so radically slowed and prolonged.

In this historical unfolding, the development of polyphony once again parallels the overflowing of the One into the Many in Neoplatonism, the One of chant bifurcating into the two voices of early organum, at first strictly parallel and then more freely independent. Indeed, some Pythagoreans considered the Dyad not truly a number (which really began, for them, with the Triad) but an all-important intermediate between One and Many. Such considerations of the status and significance of the "indeterminate Dyad" reverberate throughout Plato's dialogues and were an important part of his "unwritten

Figure 3.4
Perotin, *Viderunt omnes,* measures 1–40 (♪ sound example 3.3); the text is the same as in figure 3.3.

teachings," which identified the form of the Good with the One.[8] The contrast between Leonin's dyad (a single organal voice above the chant) and Perotin's trio of organal voices seems to reflect this primal numerology. The unfolding history of these compositions grounds the relation of the successive voices to the One in just the way that the One overflows into the Many. For the Neoplatonists, the very concept of "overflowing" implied that the One is not static or self-contained but seems to pour forth states of higher multiplicity that still bear the mark of the One insofar as we grasp their unity: the three upper voices of Perotin seem to be emanations of the chant voice below them, so that all four voices really are an emanation of the One.

Though we confront these works as written compositions, Anne Maria Busse Berger has argued that much medieval ecclesiastical polyphony was improvised according to careful formulas committed to memory, each prescribing possible counterpoints to a given intervallic motion in the underlying chant.[9] Craig Wright also deduced that the cathedral of Notre Dame was "among the darkest of the great Gothic cathedrals," the illumination so dim (and the budget for candles so limited) that most services had to be performed by heart, a practice also preferred as more devout.[10] In such cases, the written music may have been more a reminder or sample than a definitive, exclusively written text, which may have been consulted only for infrequent or exceptional occasions. If so, singers improvising polyphony confronted even more directly the difficulties of generating (as well as apprehending and coordinating) many musical lines at once. At the same time, this practice of polyphonic improvisation would probably have intensified its felt character as a collective synthesis of many independent personas into a single whole, a living demonstration of One created out of Many. The memorized yet improvised character of medieval polyphony also may have had important implications for the identities of its singers, who (as Mary Carruthers has noted) arguably did not feel themselves to be a "self" in the modern, individualistic sense but rather as "a 'subject-who-remembers,' and in this remembering also feels and thinks and judges." Likewise, Carruthers compares medieval scholars to walking anthologies (florilegia) of memorized quotations, virtually "constructed out of bits and pieces of great authors of the past."[11] Though those inner anthologies of revered quotations were generally consulted in sequence, at moments of stress or decision, the improvisation of polyphony seems to have involved combinatoric play that manipulated memorized melodic formulas according to well-practiced rules. Such improvisatory practices required and exemplified the highest degree of interplay between the personas of the several parts so that each could maintain a certain melodic integrity even while responding and harmonizing with the others.

Though organum was sung in church, after 1250 new forms of polyphonic secular music proliferated. The earliest such written work, "Sumer Is Icumen In," celebrates the coming of spring in a cheery round (figure 3.5; ♪ sound example 3.4). At the time, this was sufficiently unfamiliar that the manuscript explained how to sing it: after one singer sings the melody until reaching a red cross (after the first words, "Sumer is icumen in"),

Figure 3.5
The round (*rota*) "Sumer Is Icumen In" in manuscript, along with a modern transcription (♪ sound example 3.4). Text: "Spring has arrived, sing loudly, cuckoo. The seed is growing and the meadow is blooming and the wood is coming into leaf now. Sing, cuckoo!"

a "companion" begins while the first continues, and likewise for a third and fourth singer. These four are accompanied by two others, who each sing a continually repeated melody called a *pes*. Later called a *canon*, this technique allows a single melodic line to fold back on itself in each added voice, so that the composition is completely unified by that generating line, even though it is sung by several distinct voices.

In France, secular polyphony often took the form of *motets*. Befitting its name (from *mot*, "word"), the various voices of motets frequently had entirely different sets of words, unlike the single text underlying an organum. The thousands of motets that appeared in the thirteenth century became more and more extravagant, sometimes utilizing not only different texts but different *languages* for their different voices.[12] Consider, for example, the mid-thirteenth century motet *O Mitissima (Quant voi)—Virgo—Hec dies* (figure 3.6; ♪ sound example 3.5). Its top voice line had two different sets of words in two different manuscripts, one a Latin prayer to the Virgin Mary (beginning *O Mitissima*), the other

Figure 3.6

Motet: *O Mitissima (Quant voi)—Virgo—Hec dies* (♪ sound example 3.5). Texts: Triplum (upper line in the Bamberg Codex): "O sweetest Virgin Mary, beg thy Son to give us aid and relief against the deceiving wiles of the demons and their wickedness." Triplum (upper line in the Montpellier Codex, here included as a subsidiary line): "When I see the summer season returning and all the little birds make the woods resound, then I weep and sigh for the great desire which I have for fair Marion, who holds my heart imprisoned." Motetus (middle line): "Virgin of virgins, Light of lights, restorer of men who bore the Lord: through thee, O Mary, let grace be given as the angel announced: Thou art Virgin before and after." Tenor (lowest line): "This is the day [which the Lord hath made]."

(*Quant voi*) a French love song to Marion, a disdainful beauty who makes her lover weep and sigh. The middle line is a Latin prayer to the Virgin (*Virgo virginum*); the tenor (lowest line) is a Gregorian chant for Easter, using only its first two words (*Hec dies*). It seems that, over the tenor (functioning as the basic underlying voice, which later came to be called the *cantus firmus*) one could perform one or both of the other lines, using different sets of words or perhaps just instruments. None of the texts of these upper voices have anything to do with the text of the Gregorian tenor, except perhaps the general sense of Easter suggesting spring and love. This motet shows that such sacred and secular texts could evidently be sung together without seeming incongruous or irreverent. A well-known Gregorian chant, such as the one used here, was probably so familiar that it seemed natural to weave it into songs that probably never were sung in church.

Polyphonic singers (and their listeners) would thus have had many occasions to experience and contemplate the paradoxical relations between polyphony and the mind. In the slow unfolding of this subtle and complex issue, it will prove helpful to begin by looking back from the vantage point of the mature works of Thomas Aquinas, writing a century after Leonin wrote his *Magnus liber organi*. Born about the time that Perotin died (around 1220), Thomas was a product of the University of Paris, whose intellectual milieu literally and figuratively surrounded the School of Notre Dame and its new cathedral.[13] As the contemporary historian William of Armorica noted,

in that time [1210] letters flourished at Paris. Never before in any time or in any part of the world, whether in Athens or in Egypt, had there been such a multitude of students. The reason for this must be sought not only in the admirable beauty of Paris, but also in the special privileges which King Philip and his father before him conferred upon the scholars. In that great city the study of the trivium [grammar, logic, and rhetoric] and the quadrivium [arithmetic, geometry, music, and astronomy], of canon and civil law, as also of the science which empowers one to preserve the health of the body and cure its ills, were held in high esteem. But the crowd pressed with a special zeal about the chairs where Holy Scripture was taught, or where problems of theology were solved.[14]

This unprecedentedly large and tumultuous body of students constituted a city within the city of Paris, proudly defending their own privileges. They also heard (or even sang) the new polyphony in their new cathedral, and their attitudes and preoccupations arguably affected its reception and understanding in ways that may be comparable to the theological debates Thomas encountered after he himself arrived in Paris in 1245 as a student.[15]

We see something of the spirit and method of these debates in the "questions" Thomas used throughout his *Summa Theologiae*. Each question begins with a statement of the issue at hand (for example, "is God a person?"), to which Thomas first gives a series of objections, each in the voice of those who hold it, along with their reasoning, followed by what Thomas considers, "on the contrary," to be the decisive argument against the view held by these objections. Then Thomas responds "I answer that …," thereafter summarizing his judgment of the theological crux in light of that decisive argument, followed by replies to each objection.

Given their keen attention to these theological debates, Parisian students would likely have been struck by the new and remarkable qualities of polyphony. Indeed, the multiple conflicting "voices" consciously brought forward in a scholastic "question" themselves form a kind of intellectual polyphony. The students' thinking would have been shaped largely by Boethius's *Fundamentals of Music*, which they all studied as part of the fundamental study of the liberal arts of arithmetic, geometry, music, and astronomy.[16] Thomas too knew this text well, no less than Boethius's theological works such as his *De Trinitate*, on which the young Thomas wrote the only commentary dating from his century.[17] These early studies bore important fruit: in particular, Thomas pointedly adopted Boethius's definition of the concept of person.

In his *Summa Theologiae*, Thomas cited approvingly the argument of Augustine that "this word person of itself expresses absolutely the divine essence, as [does] this name God and this word Wise."[18] But Thomas also reviewed the subsequent controversies whether "person" really signified an *essence* or a *relation*, to which he answered that "the name person signifies relation directly, and the essence indirectly." In this way, Thomas phrased the orthodox view by saying that the Trinity comprises three persons and one hypostasis (underlying substance), which formulated both the unity and trinity essential to the triune God. The personhood of the Son *is* his sonship, in relation to the Father, whose fatherhood likewise only emerges in relation to his Son, continuing on through all the interrelations that are essential to the Trinity, where "relation" implies both the common ground of the two related persons and also their difference (between Father and Son, for instance). These relations were famously controversial; part of the schism between Eastern and Western churches concerned whether or not the Spirit could proceed from the Father *and* from the Son (as asserted in the word *filioque* inserted into the Roman Credo, its version of the Nicene Creed) or only from the Father (as most of the Eastern Orthodox churches insisted).

However deeply Thomas may have parsed the multiple personae of the godhead, the surviving evidence clearly implies that he was opposed to ecclesiastical polyphony.[19] As early as 1242, three years before Thomas began his studies in Paris, the Dominican order forbade the singing of polyphony in its houses, though there is some evidence indicating that the ban was not universally observed.[20] Having chosen (against many obstacles) to be a Dominican, rather than joining one of the older and more established orders, he subscribed to that order's pointed advocacy of chant: Thomas stated flatly that "music ought to be simple [*simplex*]," unequivocally meaning monophonic chant.[21] Yet even though he himself may have regarded critically such movements as Notre Dame polyphony, Thomas's reflections on its underlying philosophical problematic will provide a rich framework within which we can fruitfully situate his rejection of liturgical polyphony.

Thomas agreed with Aristotle that the human mind cannot think many things at once, for "it may happen that many things are known, but only one understood."[22] We cannot know a manifold "through the one form of the whole," though through "one species it [the mind] can understand at the same time; hence it is that God sees all things at the same time, because He sees all in one, that is, in His Essence. But whatever things the intellect understands under different species, it does not understand at the same time," for the same reason that "it is impossible for one and the same body at the same time to have different colors or different shapes," which in this example correspond to the "different species," the differentiating qualities through which we recognize different beings in the world.

Though he never referred directly to polyphony, Thomas was very much engaged with the closely related consideration of the differences between divine, angelic, and human minds. His questions about the nature and existence of divine and angelic beings concerned their possible interaction with humans, not merely their separate existence apart from human intelligence; hence his project required determining exactly how far human intelligence could reach. Thomas divided all created beings into those that are either purely corporeal (such as inanimate matter), those that are partly corporeal and partly spiritual (such as humans), and those that are purely spiritual, hence wholly incorporeal (such as those called "angels" in the Scriptures). As such, the angels form an interesting middle ground between purely finite, corporeal beings and the incorporeal, infinite God. Thomas argued that, as entirely actualized spiritual beings, such angels would naturally know God "by His own likeness refulgent in the angel itself."[23] In his *Treatise on Angels*, Thomas went on to explore many aspects of angelic knowledge and action. Compared to humans, who can only know God through reason, angels are far closer to the ultimate actualization of the supreme Being and hence would then know God more closely through their own natural powers.[24]

Despite this, even angelic minds cannot know the future or the secret thoughts of men.[25] Angels cannot be in several places at once, nor can several angels be in the same place at the same time; they move through space sequentially with finite speed, not instantaneously.[26] For Thomas, these deductions all follow because angels share the fundamental limitations of all finite entities. Even so, Thomas argued that an angel can understand many things at the same time. In this, the angels show their essential kinship with God, who "sees all things in one thing, which is Himself. Therefore God sees all things together, and not successively."[27] In contrast, "we understand [many things] simultaneously if we see them in some one thing; if, for instance, we understand the parts in the whole, or see different things in a mirror." Thomas took up this image of a mirror as the crux of our reflections, both literal and figurative. "A mirror and what is in it are seen by means of one species," meaning a certain mode of likeness or of shared being, in this case the species of light common to luminous objects and to the seeing eye. "But all things are seen in God as in an intelligible mirror," suggesting the paradox of a supreme mirror, whose perfect reflectivity, even apart from all objects, seems virtually to shine in the darkness.

Thomas summarized the argument we have just been considering by noting that God's own knowledge is not discursive, unlike our common modes of human cognition, which proceed through separating and delineating the strands of succession and causality in ways necessary to human understanding. God, in contrast, sees all in himself, and the angels, beholding him directly, can also see many things as one, through him. Thereby we access Thomas's resolution of the problem of polyphony: only a completely unified mind can experience many things as one, through its own unity. Only God and the angels can really comprehend polyphony because their nature is integrally polyphonic, unlike the merely monophonic human mind.

Even before Thomas's time, polyphony was controversial. In 1159, the English cleric and diplomat John of Salisbury (a close associate of Thomas à Becket) argued that "music sullies the Divine Service" on account of the "effete emotings" of the singers' "before-singing and their after-singing, their singing and their counter-singing, their in-between-singing and their ill-advised singing … to such an extent are the highest notes mixed together with the low or lowest ones. Indeed, when such practices go too far, they can more easily occasion titillation between the legs than a sense of devotion in the brain."[28] This sexual innuendo became a common complaint among those who criticized polyphony.[29]

For instance, writing at about the same time, the English Cistercian abbot Aelred of Rievaulx emphasized both the unseemliness of ornamentation and the number of voices, as if they were connected vanities, which he again connected to unmanliness:

Why that swelling and swooping of the voice? One person sings tenor, another sings duplum, yet another sings triplum. Still another ornaments and trills up and down on the melody. At one moment the voice strains, the next it wanes. First it speeds up, then it slows down with all manner of sounds. Sometimes—it is shameful to say—it is expelled like the neighing of horses, sometimes, manly strength set aside, it is constricted into the shrillness of a woman's voice. Sometimes it is turned and twisted in some sort of artful trill. Sometimes you see a man with his mouth open as if he were breathing out his last breath, not singing but threatening silence, as it were, by ridiculous interruption of the melody into snatches.[30]

Aelred here critiques specifically the contemporary compositional technique called *hocket* (*hoquetus* literally means "hiccup"), in which two voices alternately interrupt themselves by rests so as to pass their notes back and forth. His own rhetoric inflames as he mocks the bodily expression of such singers: "Now he imitates the agonies of the dying or the swooning of persons in pain. In the meantime, his whole body is violently agitated by histrionic gesticulations—contorted lips, rolling eyes, hunching shoulders—and drumming fingers keep time with every single note. And this ridiculous dissipation is called religious observance. And it is loudly claimed that where this sort of agitation is more frequent, God is more honorably served."[31] In his judgment, this has led to misleading the people: "Meanwhile, ordinary folk stand there awestruck, stupefied, marveling at the din of bellows, the humming of chimes, and the harmony of pipes," thereby including organs and bells in his censure. Nor does he think that people ignore this invitation to sensuality, for "they regard the saucy gestures of the singers and the alluring variation and dropping of the voices with considerable jeering and snickering, until you would think they had come, not to an oratory, but to a theater, not to pray, but to gawk."[32] As Christopher Page has pointed out, this controversy mirrors the conflict a contemporary poem depicted between the owl (whose sad notes lament human sins) and the nightingale, who sings the bliss of heaven.[33]

Yet the prevalence of polyphony evidenced its institutional support within the highest circles of the Church. The *Codex Calixtinus*, the first manuscript to name the composers of polyphonic sacred works, included among them four bishops (including the bishop of Chartres) and five *magistri* (masters). These were not "performer-creators" functioning as musical servitors but ranking clerics, as Wright notes, whose "eminent positions suggest that already by the mid-twelfth century, the new composer-prescriber had come to occupy a prominent position among the clerical elite."[34] In Paris about 1200, as well as in other centers, the polyphonophiles seem to have been ascendant; Bishop Odo's 1198 ordinance mandated that polyphony should be performed at the liturgies of the great feasts in Paris.[35] Such highly educated and powerful prelate-composers would surely have considered the philosophical and theological groundings of their own work, especially in a climate in which it was under attack by others within the Church.[36]

The polyphonophiles presumably would not have adopted Thomas's preference for "simple" music for human minds, while reserving polyphonic awareness only for disembodied spirits, for angels or God. Though influenced by Neoplatonism, Thomas himself stayed well back from such extreme positions as Eriugena adopted; he surely knew of Eriugena's condemnation by the Church (which occurred in 1225, when Thomas was still a child).[37] In so doing, Thomas finessed the doctrinal dangers that finally marginalized Eriugena and other extreme Neoplatonists. To do so, Thomas separated human from divine and angelic minds, consistent with his preference for "simple" human music. His use of the concept of personhood became a carefully nuanced way to express how humans share the divine image, as do angels, while still keeping them clearly separate from the divine essence so as to avoid anything smacking of pantheism.

Where Thomas advocated the Latin concept that the Trinity is three persons in one, Eriugena had adopted the Greek terminology of hypostasis and essence (discussed in box 2.2) and had thus avoided the terminology of personhood. For Eriugena, these Greek theological concepts better expressed the ultimate unity of all with God than did the more individualistic concept of personhood, which tends to imply a fundamental multiplicity (and distinguishability) of separate persons. The legal tradition seemed to emphasize the distinction between persons, each one separately individuated and represented in court, rather than the essential unity implicit in personhood as such. Thomas depicts the *mixture* (*mixtis*) of divine and human natures in Christ, rather than the Neoplatonic *blending* (*krasis*) that Eriugena advocated. Though he could perhaps have interpreted the concept of person as involving that kind of unity, Thomas chose to use personhood as a bulwark that would keep human minds (and music), however noble, from blending utterly with the divine. In contrast, the *Enchiriadis* treatises used the term *concentus* (with its implication of unified blending), rather than the milder *succentus* (implying mixture that falls short of utter union). The very overlap and possible confusion of these terms suggests how perilously close their meaning might be: the terms used to clarify the relation between polyphonic voices were, after all, the same words used to describe the interrelation between Christ's two natures as part of a single person.

Ironically, despite his stated preference for "simple" music, Thomas's own mind was startlingly polyphonic. After his youth, he customarily dictated, rather than wrote, his works. His output was staggering, amounting (on careful calculation) to more than twelve printed pages a day of closely argued prose, a pace maintained over many years. Several independent accounts by his associates confirm that, in order to do this, he "dictated at the same time on diverse subjects to three secretaries and sometimes four." Thus, Thomas's own mind seemed to have operated in a way that was distinctly polyphonic. One amanuensis even recorded that once "after dictating to him and to two other secretaries that he [Thomas] had, sitting to rest for a bit, he fell asleep and continued dictating even while sleeping."[38] Such stories of simultaneous dictation also were told of Julius Caesar to show how far his abilities exceeded the human norm. Where Caesar's prodigious feats merely indicated his superhuman talents, Thomas's ability to function polyphonically corresponds strikingly with the questions about human, angelic, and divine minds he had argued so closely, presumably while dictating to several secretaries simultaneously. Thomas's multiple dictation indicates how far a human mind could go into the polyphonic realm of angelic mentality.

Though his secretaries might have considered his feats to be miracles, Thomas himself would probably not have agreed, given his respect for what human reason and willpower could accomplish without presuming on divine grace. Thomas also knew his limitations and evidently became so weary that he would fall asleep while trying to work, as you or I might. But, unlike us, he just kept on dictating. A monk whose cell was nearby "frequently heard him talking and disputing with himself when he remained alone and without anyone else in his cell." That singular mind, capable of dictating several complex texts at once, may have been shaped by those solitary, ceaseless disputations, as if those intricate arguments in which he spoke aloud and heard himself as advocate, opponent, and judge were the crucible in which his own polyphonic mind was formed.

Writing fifty years after Thomas's death, Dante considered him one of the greatest theologians, whose thought influenced many aspects of Dante's philosophical and poetic works, particularly his *Divine Comedy*. In it, Dante expressed his admiration by giving Thomas an exalted place in his *Paradiso*, shining among a crown of "flashing lights of surpassing brightness" as Thomas generously praises his fellow theologians.[39] Yet Dante's eloquent use of polyphony seems pointedly different from Thomas's preference for "simple music." The whole *Commedia* follows a careful musical structure, as Francesco Ciabattoni has shown.[40] In the *Inferno*, music appears distorted: as the pilgrims descend deeper into hell, they hear increasing degrees of noise and cacophony, including perversions of sacred chants and such bizarre "instruments" as Malacoda making "a trumpet out of his asshole"

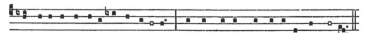

1. In éxi-tu Isra-*el de Aegý*-pto, * dómus Jácob de pópu-*lo* bárbaro :

Figure 3.7
The Gregorian chant *In exitu Israel,* sung by the pilgrims beginning the ascent of Purgatory, in *tonus peregrinus,* called the "wandering tone" because it "wanders" from the first reciting tone on A (first measure) to another on G (second measure) (♪ sound example 3.6). Text: "When Israel came out of Egypt, the house of Jacob from a strange country."

to marshal his fellow malefactors.[41] The *Purgatorio,* in contrast, begins with the strains of Gregorian chant, as repentant souls begin their ascent up the mountain singing the antiphon *In exitu Israel de Aegypto,* whose wandering "pilgrim mode (*tonus peregrinus*)" symbolizes the penitence of Jacob's children traversing the wasteland (figure 3.7; ♪ sound example 3.6). Dante emphasizes that these penitents sang in unison, "with one voice" (*ad una voce*).[42] Until the very last stages of Purgatory, the music is entirely monophonic and Dante emphasizes the distinctness and importance of the words being sung. For instance, as Matelda sings for him in the Earthly Paradise on top of the purgatorial mountain, he notes pointedly that "the sweet sound reached me *with its meaning.*"[43]

In *Paradiso,* however, the sense of the words is often lost in the surpassing beauty of the music itself, specifically as a result of the polyphonic mingling of voices. When Dante beholds the martyrs and crusaders forming a fiery cross in the heaven of Mars, he compares their singing to a consort of diverse instruments: "And as viol and harp strung [*tempra*] with many strings in their harmony will sound sweet even to one who fails to catch their tune, so from the lights that there appeared to me a melody gathered and came from the cross, enchanting me, though I could not make out the hymn."[44] Through polyphony, the blessed souls in Paradise enter into a deeper kind of knowledge than that mediated only through words and discursive reason. Part of this is political: the accord between different voices reflects and even enables the true community of the blessed souls. As the Emperor Justinian, the exemplar of justice and compiler of the Roman law, explains their concord to Dante, "hereby the living justice makes our affection so sweet within us that it can never be bent aside to any evil. Diverse voices make sweet music, so diverse ranks in our life render sweet harmony among these wheels," the revolving circles of the blessed spirits.[45] Here Dante draws on Augustine's conception (discussed above) that the "sweetest harmony" from "differing sounds" echoes the community of the saints, which is the Christian descendent of the ideal republics of Plato and Cicero.[46]

Dante here accords with such older conceptions as Hugh of St. Victor's account of the "mystic sweetness" of "different sounds in a concordant variety" imaging "the compact unity of the orderly Church." Indeed, Dante often uses the word "sweet" (*dolce* or *soave*) to describe the specific quality of polyphonic song, not merely a generic sense of pleasurable sound.[47] For him, what makes polyphony sweet is the sensation of mingled voices that reflect both the ardor of intertwined lovers and the mystic union that is God. In the heaven of Venus, he describes the descent of the blessed: "And, as one sees a spark within a flame, or discerns a [second] voice in [a] voice [*in voce voce*], when one holds [its note] while the other comes and goes, so I saw within that light still other lights."[48] The images of spark within flame, light within light, converge toward the superlative union of "voice in voice" as he describes the precise technical detail of two-voice melismatic organum, one voice holding its note while the other moves (as in figure 3.2; ♪ sound example 3.1). Dante prefers these musical evocations of polyphony because they not merely describe mystic unity through sight (in the image of seeing a spark within a flame) but allow us to feel and hear that union directly, with all its unmediated force.

Drawing on the vocabulary of dance, Dante often compares the paradisiacal life to circular dances, bringing to mind the contemporary *carole* and *rondeau*, circle dances sometimes performed on feast days even in church (see figure 3.8).[49] However controversial such dances were in some quarters (as we shall discuss in the next chapter), Dante thought they were entirely appropriate in Paradise, where they provide the choreographic equivalent of polyphony. Describing the "double dance" performed by fifteen stars, Dante notes: "There they sang not Bacchus, not Paean, but three persons in divine nature and in one person it [divine nature] and human [nature]."[50] His words can be read not only as referring to the persons in the Trinity and the person of Christ, according to Thomas, but also as describing the stars' three-part organum as three "persons" (voices) joined "in one person," making explicit the unified persona of the musical work we hypothesized earlier. Indeed, in such descriptions Thomas seems to be providing a new account that would acknowledge the "music of the spheres" as well as Aristotle's arguments that it could not be physically audible, arguments with which Thomas agreed and which presumably Dante respected. These matters reflected Dante's close interest in astronomy.[51] By having the blessed spirits in Paradise—rather than the planets or Sirens, with their pagan associations—sing the celestial music, he was able to maintain this ancient tradition while finessing the physical (and doctrinal) objections against it.

Dante's most striking description of polyphony reflects on this controversial interface between astronomy, natural philosophy, and music. It comes after Thomas finishes describing in admiring detail his fellow theologians forming a "crown of flashing lights" in the heaven of the sun. Trying to express how those many voices make together a sound "more sweet in voice than shining in aspect," Thomas seizes upon a simile that is as much auditory as visual, involving a novel and surprising device:

Then, like a clock [*orlogio*] which calls us at the hour when the Bride of God rises to sing her matins to her Bridegroom, that he may love her, in which the one part draws or drives the other, sounding *ting! ting!* [*tin tin*] with notes so sweet that the well-disposed spirit swells with love, so did I see the glorious wheel move and render voice to voice with harmony and sweetness that cannot be known except there where joy is everlasting.[52]

The kind of clock he is describing had only been invented in the West about 1271, a few decades earlier (figure 3.9).[53] Though the notion that the heavens are a mechanical

Figure 3.8
Clerics dancing a *rondeau*, from Florence, Biblioteca Medicea Laurenziana, MS. Pluteus 29.1, f. 428r.

Figure 3.9
Giovanni de' Dondi built an astrarium, a complex astronomical clock, and presented it to the duke of Milan in 1381; it disappeared sometime after 1485. This reconstruction was made by Luigi Pippa (1963); the clock shows the positions of the planets, Sun, and Moon, as well as the moveable feasts of the liturgical year.

clockwork is usually dated centuries later, Dante clearly has it in mind here, though perhaps for auditory more than astronomical reasons. He evidently has no objection to analogizing the souls in the sphere of the sun to a mechanical device, so delighted is he by the combination of exact time coordination of the many wheels (whether in the heavens or in the clockwork) "in which the one part draws or drives the other," resulting in the hourly chime, in this case signaling the hour of Matins. Thus, the polyphony of the theologians accords so closely and sweetly that their intermeshing clockwork sounds an aubade, the dawn song of lovers, here taken to signify Christ and his church.

In Dante's unitive vision, polyphony coordinates many different voices (here, the theologians forming a "crown of flashing lights" with their markedly different views) into a single delightful chiming, the true "music of the spheres." Where Thomas imputed multiple capacities of simultaneous knowledge to divine and angelic minds, Dante made explicit that those minds expressed those capacities in polyphony as the proper music for their exalted state. Although Thomas believed that humans should restrict themselves to "simple" music, Dante invited them to partake of this divine music not in the form of the legendary cosmic harmonies (which were physically inaudible) but in the rich polyphony he knew so well on Earth. Dante thus implied that to hear polyphony was to enter into a licit and divinely sanctioned paradise.

4 Oresme and the "New Song"

Despite the advocacy of Dante and others, during the fourteenth century polyphony came under continuing criticism. Its critics reiterated the accusations that polyphony was "effeminate" and inappropriate for sacred use. Such outbursts of unease reflected underlying controversies about the philosophical framework that could justify polyphonic innovations. Nevertheless, other influential voices continued to defend it, in particular the mathematician and natural philosopher Nicole Oresme. A friend to Philippe de Vitry, one of the most famous innovators of the new contrapuntal art, Oresme presented polyphony as exemplifying the "new song" described in scripture and manifest in his own far-ranging investigations into astronomy. Oresme's fundamental work in mathematics also instanced polyphony as part of his attempt to describe many-dimensional qualities in nature, including sound. To do so, he combined several interrelated variables into a single representation he called "difformity."

The highest-level condemnation of polyphonic music came from Pope John XXII, who set up his papal court in Avignon in 1316 and established there a highly centralized ecclesiastical administration. His bull *Docta sanctorum patrum* (1324) censured music in which the "melodies are broken up by hockets or robbed of their virility by *discanti* [two voices], *tripla* [three voices], *motectus* [motet], with a dangerous element produced by certain parts sung on texts in the vernacular."[1] The pope here seems to allude to the proliferation of works by Goliards or wandering singers who performed in the marketplace and in the church, where (according to a modern commentator), "under the pretext of singing motets of better quality they introduced in favor of polyphony a thousand declamations injurious to the clergy, prelates, and papacy."[2]

Going past the issue of such secular jibes, the pope's critique took a far more general turn, addressing a whole range of complex polyphony as practiced in the contemporary church:

The mere number of the notes, in these compositions, conceal from us the plainchant melody, with its simple, well-regulated rises and falls which indicate the character of the mode. These musicians run without pausing, they intoxicate the ear without satisfying it, they dramatize the text with gesture and, instead of promoting devotion, they prevent it by creating a sensuous and indecent atmosphere. Thus it was not without good reason that Boethius said: "A person who is intrinsically sensuous will delight in hearing these indecent melodies, and one who listens to them frequently will be weakened thereby and lose his virility of soul." … Therefore, after consultations with [the cardinals], We prohibit absolutely, for the future, that anyone should do such things, or others of like nature, during the Divine Office or during the Holy Sacrifice of the Mass.[3]

To defend "virility of soul," this papal edict thus connected Boethius's notion of "sensuous, indecent melodies" with polyphony, which Boethius scarcely mentioned. Yet the pope's reference to polyphony as a "common" state of things implies that it was already deeply ingrained; he exempts from his proscription "the occasional use—principally on solemn feasts at Mass and at Divine Office—of certain consonant intervals superimposed upon the simple ecclesiastical chant, provided that these harmonies are in the spirit and character of the melodies themselves, as, for instance, the consonance of the octave, the fifth, the fourth, and others of this nature," thereby explicitly including simple organum. Thus, even in the process of prohibiting much of polyphonic music, the pope still felt obliged to acknowledge that, in simple polyphony, "such consonances are pleasing to the ear and arouse devotion, and they prevent torpor among those who sing in honor of God."[4]

During the century following this edict, polyphonic music continued to flourish, even amid continuing controversy. Indeed, this particular bull may have been honored more in the breach than in the observance, promulgated by an Avignon pope considered in some quarters to have been schismatic and illegitimate, whose edicts may thus not have been widely propagated or unfailingly observed. Still, his bull long affected the French church; the records at Notre Dame in Paris, which surely had been a preeminent center of polyphonic innovation, indicate in 1408 that, in many respects, "church discant is not in use, being prohibited by statutes," presumably meaning the papal proscription then almost ninety years old.[5] Yet beginning in the early fourteenth century, learned writers compiled instructions for composing what they started to call "counterpoint" (*contrapunctus*), meaning the polyphonic texture of "note against note" (*punctus contra punctum*).[6] This new term indicated both polyphony's status as sufficiently established (or perhaps outmoded) to need such careful description and also indicated its continuing importance in music.

To assess the significance of polyphony in the mind of a particularly innovative contemporary, we will consider the provocative comments by the most brilliant figure in fourteenth-century natural philosophy, Nicole Oresme, remarks that have probably remained unnoticed because they were embedded in his mathematical works. As a prominent commentator on ancient philosophical and scientific texts, perhaps it should not be surprising that he welcomed the new music of his time, explicitly including polyphonic music. His grounds for doing so may illuminate what this style meant to him.

Oresme's larger projects went far beyond the polyphonic issues that concern us here.[7] His humble origins in Normandy did not prevent him from achieving eminence in ecclesiastical and intellectual circles. He provided new translations of Aristotle's principal scientific, ethical, and political works for Charles V, who had taken an interest in Oresme even before ascending the French throne. Along with other contemporary readers of Aristotle, such as his teacher Jean Buridan, Oresme was no slavish imitator of the ancient master; his careful and critical commentaries raised many important questions that, over the succeeding centuries, became important points of departure for later generations of natural philosophers, including (ultimately) the young Galileo Galilei. Despite long-standing narratives that depict modern science as simply rejecting ancient science, in many ways Aristotelian philosophy was the matrix within which the "new philosophy" formed; even its innovations tended to look to Aristotle for firm points of reference against which to react, especially the fundamental Aristotelian definitions and arguments that formed the background of any further developments.[8]

Nor was Oresme particularly rebellious; in his century, Aristotle had become a revered ancient source, yet not so long before he had almost been rejected as heretical and un-Christian. For Oresme, as for Thomas Aquinas, Aristotelian science represented a daring challenge to Christian thought. Oresme especially considered the sensitive question of whether time is essentially cyclical (as Aristotle and other ancients had taught) or linear, never-repeating, always new and different (as biblical revelation required). Aristotle's cyclical time also corresponded to his teaching that the world is eternal, in clear contradiction to scriptural evidence that the world had a beginning and accordingly will have an end.

Oresme's point of departure was the long-standing tension between arithmetic and geometry, though both were sisters in the quadrivium, along with music and astronomy.[9] In ancient and medieval mathematics, arithmetic concerned only whole numbers (integers) and their ratios, whereas geometry considered magnitudes that could have any size: the diagonal of a square cannot be expressed as any finite ratio of integers, as the Greeks learned early on.[10] Applied to music and astronomy, simple whole-number ratios characterized the primal musical intervals, which in turn also governed the structure and respective velocities of the heavenly bodies. If so, then it followed that the celestial motions should be cyclical: according to this reasoning, the night sky we see tonight should recur in about 36,000 years, the "Great Year."[11]

Oresme noted that, if astronomy were governed by geometric — hence incommensurable — magnitudes, then no recurrences were possible: because no ratio of finite numbers would describe the relative velocities of two such planets, even after an infinite time had passed they would never return to their initial positions. Oresme argued that such incommensurable relationships were far more probable than commensurable ones, which required perfectly precise ratios, each term exactly equal to an integer. Even more, he felt that this irrational geometric view accorded far better with Christian teaching, which disallowed endless celestial cycles: the life and death of the Redeemer happened once and once only, hence contradicting any notion of eternal return of past events into the future.

Yet in a debate he staged between personified figures of Arithmetic and Geometry, Oresme himself seems finally to have sided with Geometry, who rejects eternal repetition in the cosmos for essentially musical reasons, because it would correspond to endless repetition of the same melodies over and over, ad infinitum.

What song would please that is frequently or oft repeated? Would not such uniformity produce disgust? It surely would, for novelty is more delightful. A singer who is unable to vary musical sounds, which are infinitely variable, would no longer be thought best, but [would be taken for] a cuckoo. Now if all the celestial motions are commensurable, and if the world were eternal, the same, or similar, motions and effects would necessarily be repeated. Also, of necessity, there would be a Great Year, which, in the eyes of God, has no existence—indeed, it has even less existence than a day that has passed into yesterday. For this reason it seems more delightful and perfect—and also more appropriate to the deity—that the same event should not be repeated so often, but that new and dissimilar configurations should emerge from previous ones and always produce different effects. In this way, the far-stretching sequence of ages, which Pythagoras knew as the golden chain, would not return in a circle, but would always proceed endlessly in a straight line.[12]

Even apart from the decisive weight of scriptural evidence against the eternity of the world, Oresme confuted this ancient view simply on the evidence of musical experience: the cosmic harmonies cannot be endless repetitions, lest the cosmic work of art cause "disgust" through excessive uniformity.[13]

In his final work, *Le Livre du ciel et du monde* (1377), Oresme expanded this critique in the course of his commentary on Aristotle's *On the Heavens*. Reiterating and rephrasing his earlier conclusions, he notes that the heavenly bodies "are continuously producing new but imperceptible music: *canticum novum*, a new song, such as never existed before."[14] This "new song," as we saw, is the hallmark of his musical argument for incommensurable motion, according also with Scripture, which "often speaks of the divine music of the angels and blessed souls caused by God Himself: They were singing a new canticle [*canticum novum*] before the throne."[15] Oresme specifically praised the polyphonic quality of this new song:

Since the bodies of our world are governed by heavenly bodies and by their natural movements, as Aristotle says in the first book of *Meteors*, it follows therefore that terrestrial bodies are continuously in new and different arrangements such as never previously existed and that human affairs, except those that depend upon the will as opposed to natural inclination, are continuously different and such as they never were before in any way at all. Just as change cannot exist unless it is for better or worse—although both better and worse are sometimes for the best—and just as choral singing by excellent voices is not so good if the voices always sing in absolute harmony [*tousjours ou tres meilleur acort*], in the same way things here below are sometimes in better state than at other times, depending upon the variations in the imperceptible music of the spheres; accordingly, sometimes we have peace, sometimes war, as the Scripture says: *Tempus belli, et tempus pacis* [a time for war and a time for peace]; one time sterility, another time fertility, and so on with all the other changes.[16]

Such an explicitly philosophical reference to polyphonic singing was rare during the Middle Ages. Oresme praised the "song of many voices" (*chant de pluseurs voiez*) as having more variation than did the same excellent voices "always or very well in accord." This changing many-voiced song reflects a cosmos whose elements are "continuously different and such as they never were before in any way at all." The course of cosmic and human history is like a polyphonic composition, in which the "song of many voices" departs from perfect consonance precisely in order to bring about even greater felicity hereafter. Oresme thus asserts that the possibilities of variation in monophonic music are markedly less than those for polyphony, not only in terms of quantity but in quality as well: the felt impression of variation of many voices singing not in unison is incomparably different from that possible in a single melodic line, as varied as it may be.

Oresme's account also addressed a troubling discrepancy: though the incommensurable celestial motions never repeat, cosmic music (and human history) should have an ending that is the musical image of the Last Judgment, the end of time presaged by revelation. In effect, Oresme revised the endless (and unendable) unfolding of the harmony of the (incommensurable) spheres by imposing on it the form of a huge polyphonic motet, moving irreversibly from its beginning, through manifold dissonances and interplay between voices, to a consonant close. Such a grand culmination would be in deep tension with the intrinsic infinitude of incommensurable motions, which may have troubled Oresme, though he never commented on it.[17]

Oresme's praise of variation and of polyphonic music was unequivocal and remarkable. As a ranking ecclesiastic, Oresme doubtless knew of the papal bull condemning polyphonic music (however much or little that edict was actually obeyed in various dioceses). He must also have been aware of the aversion of the Dominicans and other religious orders to polyphony. His justification of polyphonic diversity, then, offered a kind of sonic counterpart to Abbot Suger's praise of visual opulence as part of divine worship: "the dull soul is led to the vision of God by beautiful things."[18]

The crux lies in Oresme's allusion to the *canticum novum*, the "new song" he adduced from many scriptural sources, which illuminate the precise force of this phrase. King David's Psalter speaks several times of the "new song" the soul should offer the Lord, as an expression of hope and renewal because of God's grace. The Book of Revelation takes this image even further: after the climactic opening of the Book of Seven Seals, those present "fell down before the Lamb" and, with their harps and vials of precious odors "which are the prayers of saints," offered the "new canticle [*canticum novum*]"; "Thou art worthy, O Lord, to take the book, and to open the seals thereof; because thou wast slain, and hast redeemed us to God, in thy blood, out of every tribe, and tongue, and people, and nation. And hast made us to our God a kingdom and priests, and we shall reign on the earth."[19] This new song emerges at a truly apocalyptic moment and ratifies the Lord's actions to end the world and time itself. Its singers reveal the new song as signaling the "new heaven and new earth" that shall emerge after the destruction of the old.

This unequivocal reference thus tells us that polyphony, as Oresme portrayed it, gives us entry to the New Jerusalem, the promised paradise beyond the tears and suffering of history. Polyphony may thus constitute an entrance to paradise itself, experienced musically. The singers of the new song form a whole chorus, comprising "the four living creatures" and "the four and twenty ancients" with their harps.[20] Oresme clearly took this to be a scene of polyphonic music par excellence, perhaps its primal defining moment. Even the multiple sweet odors that accompany the new song seem to be a polyvalent distillate of the diverse "prayers of saints." Later in the Apocalypse, 144,000 virgins, "harping on their harps, … sung as it were a new canticle [*quasi canticum novum*], before the throne," in the presence of the original twenty-eight singers.[21] The narrator informs us that "no man could say the canticle" except those elect virgins "who follow the Lamb whithersoever he goeth … purchased from among men, the first fruits to God and to the Lamb … for they are without spot before the throne of God." Thus, Oresme's emphasis on these two parallel passages connected the origins of polyphony with its dissemination among the spiritual elect.

The new song also harks back to numerous passages in scripture whose manifold resonance has a kind of polyphonic force. "Behold, I make all things new," says the Lamb upon the throne during this scene from the Revelation, calling to mind (among many other passages) Paul's declaration that "if there any be in Christ a new creature, the old things are passed away, behold all things have become new."[22] The essential act of baptism or salvation is a new birth, a new song. Thus, Oresme's encomium of variation is fundamentally a reflection on the essential newness of divine grace, which infuses life and song. He notes that "accordingly, it could well be that these heavenly bodies, being constantly in new positional relationships never previously experienced, are continuously producing new but imperceptible music [like that mentioned in Holy Scripture as] *canticum novum*, a new song, such as never existed before [*si que onques ne fu tel*]." *Such as never existed before*: this song is not simply a new version of an old song but is a radically new kind of singing, as different as polyphony is from monophony. The ever-changing face of the heavens and the intricate variability of polyphonic song are not merely a kaleidoscopic play of combinatoric variations but the living shape of divine activity made perceptible through astronomy and music. Oresme's God makes his presence felt through the endless, unrepeatable flow of new melodic life both in each voice (each separate being) and in their polyphonic interweaving, which gives a further dimension to their lives.

Oresme goes on to say that "we could perhaps say that the heavens are like a man who sings a melody and at the same time dances, thus making music in both ways [*fait double musique*]—*cantu et gestu*—in song and in action."[23] This "double music" shows astronomy as inherently polyphonic, because it unites the pure song of mathematical quantities with the "dance" of the visible planets: Oresme refers to a *carole*, a circle dance of his time much criticized by contemporary preachers (figure 4.1).[24] His use of the imagery of dance directs attention to the *rhythmic* aspect of music, rather than only pitch relations. Oresme describes the patterning of cosmic motions in terms of the rhythmic microcosm of human dance.

Figure 4.1
An illustration of a *carole*, the controversial round dance, from a manuscript of *Roman de la rose* by Guillaume de Lorris and Jean de Meung, dating from 1401–50 (Bibliothèque nationale de France, Ms. fr. 1665, f. 7r).

Oresme earlier had argued that what he calls divine music "is experienced by angels and souls of the blessed" but (following Aristotle) is not audible, in contrast to our "human music" of voices and instruments. Between these two lie what Oresme calls *musica mundana*, the music of the whole cosmos (*mundus*). By comparison with the divine realm of the "new song," Oresme notes that "our world [*le monde*, the cosmos] makes only the music of action, like the man dancing and following the rhythm of the song to which he is listening. This may possibly be the meaning of St. Augustine, Boethius, and those others who call mundane [*mondaine* in the sense of cosmic] music celestial. it might have spread to bodies here below in the manner just described."[25] We come to know this cosmic music "by human reason," lacking the direct perceptivity possessed by angels and the blessed. Oresme thus revises the Boethian threefold division between *musica mundana*, *musica humana*, and *musica instrumentalis*, which lacked "divine music" and had two kinds of human music, that of the human body as a microcosm (*musica humana*) and instrumental music (*musica instrumentalis*).

By making this change, Oresme emphasized the transcendent aspects of "the divine music of the angels and the blessed souls caused by God Himself," which is the archetypal new song, citing the passage from Revelation we have discussed in detail. What had been the highest level of music for Boethius was for Oresme (and others before him in the Christian tradition) only second highest, the "music of the spheres" intelligible to human reason. But because of the parallelisms and connections between these levels, we learn that the new song of divine music is the pattern for the music of the spheres, whose polyphony and incommensurability are mandated by the highest levels of divine grace. Similar parallelisms extend down to the level of human music because "our world [*le monde*] makes only the music of action, like the man dancing and following the rhythm of the song to which he is listening." In so doing, audible human music "conforms to and has a close relationship and affinity with heavenly music," dancing to rhythms that ultimately have their origin in heaven.[26]

Elsewhere in his mathematical work, Oresme also used sound and music as important examples of what he considered the polyvalent aspects of all observable quantities. Here, he took up the work begun by the "Oxford Calculators" in preceding decades, especially Thomas Bradwardine, the "*doctor profundus*" who served briefly as Archbishop of Canterbury toward the end of his life. Like other contemporary students of Aristotle, these Oxford scholars reexamined the place of quantity and quality in his natural philosophy, even though Aristotle himself had argued that the mathematics of eternal forms (arithmetic and geometry) could not describe the variable physical world of growth and change. Yet the question remained: exactly what can be said about the various properties of a changing substance? To what extent can the precise language of quantity, not only of quality, be used to describe degrees of whiteness, of heat, of speed, of loudness? In addressing such questions, Oresme found ways to represent in a single graphic image the interrelation of several different variables, the mathematical equivalent of grasping many voices as one.

In his *Tractatus de proportionibus* (*Treatise on Proportions*, 1328), Bradwardine generalized the theory of proportions found in Euclid's *Elements* to address many such questions. Aristotle had stated that the speed of an object traversing a resisting medium is proportional directly to the moving force and inversely to the resistance, but if resistance exceeds the moving force, this proportion would fail because the object could not move. To deal with such problems, Bradwardine developed more general proportions that go (as we would put it) as the nth power or nth root of some quantity. He used these to describe what we would call exponential growth, in which a quantity increases in proportion to its value at any time, just as compound interest grows in proportion to the current value of its principal. Bradwardine and his Oxford colleagues also were able to state what we call the *mean speed theorem*: A body moving with constant speed travels the same distance as one constantly accelerated that eventually reaches twice that speed. This is essentially the Galilean law of uniformly accelerated motion, though the Oxford school did not offer any experiments to test its validity. In his *De causa Dei contra Pelagium* (*On the Cause of God against the Pelagians*, circulated in manuscript but not published until 1618), Bradwardine even argued—against Aristotle's denial of a vacuum—that space was infinite and void, possibly containing other worlds. Oresme further generalized important aspects of Bradwardine's work, taking the concept of exponents and roots to include irrational quantities, not only integers and fractions. This was an important element of Oresme's program to study the relationship between commensurable (rational) and incommensurable (irrational) motions in the physical and astronomical realms.

In the course of his *Tractatus de configurationibus qualitatem et motuum* (*Treatise on the Configurations of Qualities and Motions*, about 1351–55), Oresme discussed the possible interrelationships between the qualities of a single substance over time; his particular innovation was to give a visual, imaginary representation of the changing "intensity" of the given quality in a form one might compare to modern graphs, down to the perpendicular orientation of our Cartesian coordinates:

Therefore, every intensity which can be acquired successively ought to be imagined by a straight line perpendicularly erected on some point of the space or subject of the intensible thing, e.g., a quality. … For whatever ratio is found to exist between intensity and intensity, in relating intensities of the same kind, a similar ratio is found to exist between line and line. … Therefore, the measure of intensities can be fittingly imagined as the measure of lines, since an intensity could be imagined as being infinitely decreased or infinitely increased in the same way as a line.[27]

We are so used to such graphs that we should not take them for granted; in Oresme's description, we can see his imagination groping to give visual representation (which he calls a "configuration") of inherently nonvisual qualities, such as swiftness or loudness, as well as of such visual qualities as whiteness. In so doing, he acknowledges that this "configuration" cannot be naively understood to exist in perceptual space: "Of course, the line of intensity of which we have just spoken is not actually [*secundam rem*] extended outside of the point or subject but is only so extended in the imagination [*ymaginatur*]," so that the choice of perpendicular graphing is only a "more fitting" imaginary choice of representation, not a literal depiction.[28] In making this convention, among the salient "dimensions" (as he calls them) Oresme calls the extension in time the "longitude" and the intensity of the given quality at that time its "latitude," as if locating them on a map, though long before modern cartography and the present meaning of these coordinates

His simplest example is what he calls a "uniform configuration," meaning an intensity that is constant (such as a sound whose loudness does not change), which he represents as a rectangle, its base representing time, its height the constant intensity. Oresme goes on to consider more general kinds of "difform configurations," meaning those that are irregular or asymmetric (from the Latin *dis-formis*). For instance, he calls a "uniformly difform configuration" an intensity that changes at a constant rate over time, which he represents as a right triangle, suitable to represent a sound constantly growing in loudness or a body moving with constant speed.[29] He interprets the area of this diagram to represent the distance traveled by such a body (for instance), adding a further level of interpretive meaning we can compare to modern graphs (figure 4.2).

Oresme also considers more than two intensities operating simultaneously, up to and even beyond four of them, thus involving what he calls the "fourth dimension," perhaps the first moment in Western thought in which this possibility had been described using the terminology of "dimensions." To be sure, he immediately notes that "it does not happen that a fourth dimension exists or is imagined," but even opening up such a notional

z

diffo:mis vnifo:miter variatio reddit vnifor
mi:ter oiffo:mi:er oiffo:mes. ¶ Latin: vni
forin i oii:ofis c illa q̃ inf excellus graduu3
eq̃ oiftáuu3 fuat caidé pportó3 aid tñ a p
portóe eq̃litatis. Ná a uit excellus graduu3
inter ie eq̃ oiftuiiuu fuarent pportó3 eq̃ta
tis .uc cci .antu: vnifo:mic oiucfis ut p3 ex
oiffuuuonibus membrorum fecúde oiuifiois
Rurfus ii nulla proporcio feruat tunc nulla
poffet attendi vnifoimitas in latitndine tali z
fic non effet vnifoimiter oifio:m i oiffo:mis
¶ Latitu: oiffoimiter oiffoimiter oiffo:mis
c illa q̃ inter excellus graduu eq̃ oiftantiu3
non ieruat eandem proportionem ficu.?in fe
cunda parte patebit. Tlorandum tame:i eft
cp ficut in fupradictis oiffuiitóib⁹ ubi logtur
oe excellu graduum inter fe eq̃ oiftantium
oeb3 accipi oiftanc'a fcóm partes latitudinis
excéliue z nó intéfiue via ut loquunt o.c:e oif
fin:tóe3 ó oiftátia 5 duii fituali ii aut graduali

oiffóric oiffofis

ot) oi, oiitofis

Equic fcóa ps in qua ut
fupradicta intelligatur ad
fenfu3 per figuras gcome
tricas oftenduntur. Et ut
omnem fpeciem latitudis
in prefenti materia via oc
currat apparentior i tinudies ad figuras gco
metcas a ⁚plicant. Ista ps oiuidif p tria ca
pitula q̃u⸗ p᷑ ᷑tinet oiónce.z᷁ fuppofiió⸗s

Figure 4.2
Diagrams of different kinds of difformity, in which the horizontal axis represents time and the vertical represents degrees of velocity, from a text presenting Oresme's ideas, the *Tractatus de latitudinibus formarum* (published 1482, formerly attributed to Oresme but probably by a student of his).

possibility was momentous: he identified "dimension" with some parameter that is not necessarily spatial.[30]

Even more important for our purposes, Oresme thereby stitches together a multiplicity of magnitudes into a single unified configuration. In so doing, he provides a conceptual framework that helps us understand how he phrased and understood the problem of polyphony, the interrelation of many voices in a single composition. Indeed, he explicitly applies his theory to sound, as well as to other qualities, such as whiteness and velocity. His application to the description of varying velocity bears comparison with Galileo's much later work in this vein, but his work on sound has not been so much remarked, perhaps because the underlying issue of polyphony needed to be made explicit.

When Oresme turns to sound, he immediately connects it with a musical problem Aristotle and Boethius had already raised, namely that "there is in sensible sound a certain discreteness brought about by the interposition of pauses which sometimes are so frequent and so small that they are not perceived by the ear and the whole seems to be one continuous sound."[31] Oresme instances the intermittent, discontinuous quality of sound heard from "a large trumpet or great tube," the sputtering caused by the lip vibrations, the intermittent puffs of air needed to excite the instrument to speak.

He thus emphasizes the discrete aspect of sound, familiar to us because the images we see in movies or computers are in fact also intermittent, needing to be refreshed about twenty-four times per second in order to seem continuous. Curiously, though this is for us a fairly familiar concept regarding visual images, it is far less familiar in relation to musical sounds, which tend to be treated as if they were completely continuous, perhaps because (unlike movies) the sound-producing mechanisms that trick our ear are so ancient that we have forgotten their inherent discontinuity. We do not, after all, usually perceive the intermittency of the sound of the bow on the violin string, unless perhaps the player produces that curiously grating sound as the bow seeks the right velocity to excite the string sufficiently smoothly. As we shall see, only in the nineteenth century did Hermann von Helmholtz and others analyze the rapid alternation of grip and release that constitutes the action of the bow on the violin string.

Oresme was, however, familiar with the intermittency of sound and therefore raises the question of in what sense a rapid sequence of many sound pulses is really *one*, because

that sound which is interrupted by sensible pauses is said to be one improperly and [only] by aggregation. And indeed this aggregation is of two kinds. One is simple aggregation in which there are not several sounds existing simultaneously as in a single cantilena or a single antiphon. The other is a composite aggregation in which several sounds resound simultaneously, as takes place when joyful choirs mingle pleasant melodies. So sound is said to be one sound in four ways.[32]

Here Oresme explicitly raises the problem of polyphony, both connecting and distinguishing it from the problem of aggregating sounds sequentially in time to form a single melodic line; he contrasts polyphony against cantilena or antiphon, each a kind of monophonic song. He also notes that "sound, as well as motion has two kinds of extension," loudness and pitch, so that sound can have many sorts of difformity, uniform as well as nonuniform. Sound can be beautiful or ugly in many possible ways, which he sets about analyzing and enumerating systematically.

In general, Oresme's criteria for beauty of sound involve a subtle mingling of unity and variety, tending to a temperate mean that recalls Aristotle's advocacy of the ethical mean as guide for the good human life and also reminds us that Oresme had translated Aristotle's

Politics for Charles V.[33] Though Aristotle famously recommended the mean, he also emphasized that some qualities (especially courage) should be at their maximum, while others (such as desire) should be moderated in ways that respond to circumstance and judgment. Oresme considers that a perfectly continuous sound (were such possible) would be beautiful if it had "temperate pitch," "moderate strength," "uniform pitch," and "beautiful difformity in strength [volume]," for even had it only the first three qualities, "experience teaches that such a sound" would be "rendered more delightful by duly increasing or decreasing its strength."[34] Thus, a pure pitch by itself, however temperate and uniform, would be more beautiful were it to swell or diminish artfully in volume.

If a sound seems continuous but is really intermittent, Oresme calls for careful regulation of "the duly measured quantity of the imperceptible, small sounds. For if the [interrupting] pauses were excessively long although the whole sound seems to be a continuous one, or even if the small sounds interrupted each other or were improperly measured or badly proportioned with respect to the pauses, then the sound will appear raucous or rough or vitiated by a certain note of ugliness."[35] Here Oresme may have given one of the earliest discussions of timbre as an important and distinct aspect of sound, another aspect of its essentially polymodal complexity comprising many independent qualities, not just volume, pitch, and duration. He recommends a certain "tempered roughness" that will avoid the annoying sound of "certain voices which are judged to be insipid because of their excessive lightness and a quality of continuousness, as perhaps the cry of a cat and the voice of certain men who seem to have windpipes that are smoothed or oiled."[36] In contrast, "tempered roughness" avoids unctuousness precisely through the artful admixture of sound and silence; the beautiful unity of the voice plays off against its underlying diversity (even if only indirectly perceived).

To effect this, Oresme requires "a symphonic mixing [*simphonica conmixtio*] [in the succession] of the small sounds," following the harmonic ratios, in which one term in the ratio is drawn from the series of successive triples, 1, 3, 9, 27, 81, …, and the other from the doubles 1, 2, 4, 8, 16, …, thus yielding the familiar musical ratios such as 1:2 (octave), 2:3 (fifth), 3:4 (fourth), and 8:9 (whole tone) that Oresme knew from Boethius and from Plato's *Timaeus*. Though these ratios were commonplaces of musical theory, here Oresme offers a new insight into their exact function, which lacked precedent in medieval or ancient musical theory: for him, the ratios regulate not just the respective string lengths but intrinsically govern the "fine structure" of sound and silence within the perceived sound of the interval, heard as *simultaneous*, not successive, sounds, so that harmonic ratios describe the "temperate roughness" of two voices blending simultaneously into a single perceived interval. Note also that by using the term "symphonic" for this blending, Oresme confirms its specific current meaning, which we have already seen since the ninth century in texts describing polyphony.

Oresme also notes that if the ratio of string lengths is not harmonic, especially if it is "an irrational ratio," the sound will be "excessively ugly." Following Book X of Euclid's

Elements, Oresme notes that there are degrees of irrationality, so that "certain ratios are further removed from, and more extraneous to, harmonic or consonant ratios than are others, and accordingly the sound can become uglier and uglier."[37] Thus, though such "irrational ratios" have an important and respected place in geometry (characterizing, for instance, the diagonal of a square with respect to its side), they are audibly ugly, to the degree of their relative irrationality (♪ sound example 4.1).[38] By implication, the *auditory* perception of beauty is governed by rational intervals (such as the octave, 1:2, or fifth, 2:3), though *visual* perception might admire the beauty of irrational proportions, such as the diagonal of a square. There are points of commonality between the auditory and the visual; Oresme notes that

in the mixture of colors by imperceptible parts, if they are duly proportioned as to quantity and intensity, the whole will become beautiful and if unduly proportioned the whole will be ugly, as is evident in the mixture of wools in cloths; or also if a top which is of two or more colors were spun very quickly, then if the simple colors are well proportioned in quantity and intensity a beautiful mean color will appear. But if they are not well proportioned, an ugly mean color will appear. The same thing is true of a sound which appears to be a single sound but is a mixture of partial sounds,

for which he instances the sounds of bells or strings, which he says can sound "false" or "true."[39] Thus, the phenomena of well- or ill-proportioned mixtures are common to sound and vision, a comparison that helps us see the ways in which Oresme embraces multiplicity in both realms, as if the comparison between them indicated ways in which the ear, like the eye, could "scan" and blend discrete elements into a single whole.

Oresme further addresses the unity "of several sounds which succeed each other after the interposition of pauses, as in an antiphon or a certain kind of cantilena," meaning a monophonic song that has measured rests, not just a single stream of melody without pause. This example refers to the new art of measured music, later called *ars nova*, whose master, Philippe de Vitry, was an admired older contemporary to whom Oresme had dedicated a book and who shared his musical and mathematical interests (as I have discussed elsewhere).[40] Along with Jehan de Murs (on whose work Oresme also drew), de Vitry provided a newly sophisticated notation to allow the precise designation of rhythms, especially for the needs of polyphonic music, of which he was a leading composer as well as theorist. More subtle but no less important, de Murs and de Vitry also found new ways to measure rests, not just sung notes, enabling notation that could coordinate ever more complex polyphonic music.[41] Oresme may have been one of the few contemporary commentators who emphasized the profound effect of being able to measure silence no less than sound, which he first considers in the context of a single voice. In what way is a melody "one" if it includes pauses that might tempt one to think that the song has ended? If, he notes, a song included a pause of one hour we would scarcely consider such a song "one"; "thus a pause is called 'long' which lasts a notably long time and annuls every kind of unity of sound."[42]

With that in mind, Oresme turns to the various durations of "short" pauses that would not disrupt the unity of the song but whose presence, in fact, can introduce a new sort of beauty. He recognizes that, as such pauses become minimal and imperceptible, they go over into the minute interruptions discussed above that constitute an unnoticed but important part of seemingly continuous sounds. But as the pauses become perceptible rests, then Oresme argues that they should be "duly measured," so that the sound not be "too minutely, or even too grossly, cut up and divided unproportionally." As with his earlier praise of mean proportion, Oresme recommends "a due harmonic difformity according to pitch, or a fitting harmonic mixture" that will respect issues of pitch as well as of rhythm in an integrated whole.[43]

In so doing, Oresme seems to recommend the new rhythmic notation, which is organized by arithmetic proportion, just as our rhythmic notation continues to be; in de Vitry's notation, rests, like sounded notes, are divided proportionally in units of 2 or 3, so that thereby the interrelation between notes and rests will inherently remain proportionate and use the same integral ratios that also characterize the harmonic pitch-relations discussed above: 1:2, 2:3, and so forth. Arithmetic proportion thus governs pitch and rhythm, sound and silence, giving a kind of unity to melodic lines now capable of articulation and interruption. The unity of each melodic line finally rests on the ways number governs all aspects of the music, which grounds and limits the "due variation" that Oresme also recommends.

This same unity also underlies Oresme's final case, "the composite aggregation of many simultaneously resonant sounds," meaning polyphony in its full generality, in which each voice involves imperceptible as well as perceptible interruptions and pauses. He not only admits but encourages the multiplicity of voices because "two sounds do not produce as good a consonance as three or more sounds," though, carried too far, "there can be so many sounds that a certain confusion results." Despite this possibility, he prefers "due variation" and the attendant richness that a greater multiplicity of voices gives to their joint resonance. He emphasizes the "consonance or symphony of the sounds, i.e. their sounding together simultaneously, and this in a melodious fashion, for otherwise there would be dissonance or discord."[44]

The simultaneity of polyphony is the essence of its "symphony," tempered always by the underlying use of "harmonic ratios" and harmonious metrical proportions of sound and silence. For the first time in his discourse about musical beauty, he speaks of "a certain delightful and decorous order, as good composers of songs have known": polyphony may be capable of eliciting a kind of delight that goes beyond the beauties of monophony by introducing a new degree of "difformity," of interconnected variation. Though he always seeks a certain mean between extremes, Oresme wishes to increase variation: a polyphonic song that is uniform in volume is less pleasing to him than one that involves an infinite range of increase … [including] increasing intensity of the volume and high or low pitch and similarly from the figuration of difformity in pitch and volume which also can recede by any amount toward infinite from a proper and harmonic configuration. But if a sound has some of the good circumstances and is deficient in the others, then the sound will not be purely beautiful nor purely ugly but will be a mean and this mean can be varied in many ways and has a great range according as the good circumstances are many or few, more or less fundamental, more or less intense.[45]

In this context, Oresme intends the full complexity of the Aristotelian mean, for the "absolute beauty" he describes "has all the good or beautiful conditions" to a superlative degree. As with the ethical quality of courage, it is not possible to be too beautiful or too good, to which end Oresme advocates increasing diversity, variety, and intensity as much as possible, consistent with the underlying harmonic proportions.

For him, polyphony offers the highest possibilities of sonic beauty precisely because it opens other novel avenues for all these qualities and is thus quintessentially "new" in this positive sense of inviting and even generating possibilities for greater artistic effect. Though not part of the inherent harmonic structure of absolute beauty, and hence "accidental" in the technical Aristotelian sense, Oresme praises "the unaccustomedness in hearing such a beautiful sound. For sometimes, from this unfamiliarity and novelty, admiration is produced and this admiration produces delight [*delectationem*]. And from this delight the sound is judged by the man unaccustomed to hear such sounds to be more beautiful."[46] Novelty causes delight through wonder and admiration, which is the recognition of the wonderful: *ad-mirari*, to wonder at something. In that sense, we realize that "delight" or delectation augments the apprehension of beauty through relishing its newness.

To explain this, Oresme brings up another "accidental" circumstance, namely

the memory of things past, so that if anyone is very happy and at the time hears a beautiful melody so many times that a semblance of the melody is impressed in his memory, then it happens afterwards, when he hears a similar sound or song, that then concurrently his present memory of pleasure returns to the state in which it was while he first listened to the song and so accordingly he takes greater delight [in it] unless some present unhappiness stands in the way. … Hence it is that we see old men commonly take delight in songs which they learned in the time of their youth when evidently they were more fresh and cheerful than they are now.[47]

The principle of delight is explicitly connected with freshness and youth; the "new song" is perpetually the song of one's youth and hence also the song that makes one young again, if only for a time. In contrast, those whose senses "are so slow and dull or obtuse and judgment so slow that they cannot easily appreciate the subtle and beautiful variations and quick beats in sounds … like old persons or youths who are dull by nature" cannot delight in the new because they are incapable of apprehending it. "But other people—such as youths of a sensitive nature—for the opposite reason take great delight in such [subtleties]."[48] This praise of youth, novelty, and delight to some extent goes against Aristotle's well-known definition of happiness (*eudaimonia*) as a ripe quality discernible only near life's end, a quality he thought unattainable for young men, with all their intemperate rashness.

Oresme may here be informed by biblical insights that speak against Aristotle, who, though a revered authority, was still a pagan ignorant of sacred Scripture. When he assembles his praise of the power of music, Oresme adds to the classical sources (such as the story of Orpheus) by arguing that the highest perfection of music will be known "in another age," after the Apocalypse in which the new song will finally be heard as it never was before. Yet again referring to this biblical passage, Oresme notes that this new song deserves its name "because of the continuous innovation in the figuration of this difformity in sound without a disagreeable repetition of one and the same [figuration]."[49] As the damned will suffer from hearing "a certain difform difformity, gloomy, deprived of every good circumstance, offending and saddening the miserable listeners," the positive difformity of the new song will delight the blessed endlessly. The essence of their musical felicity is the difformity of the new song, its multiplicity, its never-ending flow of delightful innovation, mirroring the eternity they are enjoying, which is not an endless, featureless time but a continual outpouring of ever-new beauty and wonder.

Turning from celestial prophecy, Oresme finally connects "the difformity of sounds to the magical arts" as part of his polemic against the pretensions of magic and sorcery. Oresme's unsparing attack on what he calls the "most fraudulent" claims of magic and astrology relied both on natural philosophy and the evidence of scripture. He dismisses the theory that magic operates through the conjuration of demons, whom he considers to operate only as "permitted by divinity" and thus cannot be commanded by men in ways "alien to natural philosophy and to true doctrine."[50]

Those "falsely called necromancers" gain much of their power from "the lying persuasion of that which is false" operating on "stupid people," especially "boys and adolescents or youths, because of weakness of mind and easy credulity" or "certain old women whose imaginative power has been vitiated and corrupted."[51] But there remain modes whereby music can operate in ways that the ignorant might term magical or that could even fall under what were commonly considered natural affinities. For instance, Oresme considers the widespread belief that "strings made from wolf gut can never harmonize or accord with strings made from sheep gut" or that the sound of a drum made from wolf hide will destroy drums made from sheep skin, presumably as a result of natural antipathies between these species. Ignoring this quasi-magical possibility, he hypothesizes that perhaps the wolf string or drum is inharmonious with itself or with the sheep-made instrument purely as a result of natural causes, such as conflicting resonances between the difformities in pitch and loudness of the two sounding bodies.[52]

Thus, Oresme argued that what is commonly referred to as the powers of magic actually emerge from intelligible and natural causes, of which music is a particularly potent example. If "ordinary music" has the kind of power known since ancient times, "it follows that some special and strange difformity or configuration of sounds could be devised that would have greater and more marvelous effects, just as we see that by the art of medicine compounds of marvelous power (like theriac [a legendary antidote and panacea] or some such compound) are made. And so it is that the magical arts are based in part on the power and force of a certain configuration of sounds, both in melody and in words."[53] It follows that the art of devising such "special and strange difformity" might have all the force attributed to magic by the ignorant, though these wonderful powers would be gained purely through natural sources. The mastery of sound, in all its polyphonic possibilities, would then be the only legitimate and authentic magic, granting its adepts powers that Oresme analyzed and defended, though he dismissed the rest of necromancy.

5 Polyphonic Controversies

During the fifteenth century, musical controversies mirrored the growth of reformist attitudes. The ensuing polemics led to what Rob C. Wegman called the "crisis of modern music, 1470–1530," a controversy about polyphony in which entered important elements of the nascent Protestant Reformation, along with several long-simmering polemic issues we have already encountered.[1] Reformist critics who emphasized the intelligibility of sacred texts and deprecated complex musical artifice found polyphony an ideal target. These musical controversies at one point even led to armored combat. Though such figures as John Wyclif, Girolamo Savonarola, and Desiderius Erasmus attacked polyphony, other voices rose in its defense, including Leonardo da Vinci, who argued that through polyphony music rose to its greatest intellectual and artistic significance, able to excel poetry and even to rival painting.

As we have seen, the papal ban was part of a larger current of critical reactions to polyphony. Many of these concerned the practices of religious orders in primarily monastic settings. The Dominican ban on polyphony (1242) was followed by similar proscriptions by the Augustinians, Benedictines, Carthusians, and Cistercians.[2] These bans shared the critical language of John of Salisbury and the papal bull; in a 1380 Carthusian treatise on plainchant, the author noted that he could not recommend polyphony (*discantus*) "because I am a monk, and must not give opportunity for lasciviousness."[3] Still, this author spoke favorably about *simplex harmonia*, a kind of simple organum (plainchant accompanied by parallel fifths and octaves). Thus, their objection seemed to be more against elaborate rhythms and melodies than several voices as such. Other authors shared these views, including Thomas à Kempis, whose famous book *The Imitation of Christ* (written about 1420) popularized the humble and sincere piety of the reformist movement called "modern devotion" (*devotio moderna*). He too endorsed a simple kind of polyphonic singing (*organum simplex*), as opposed to the objectionable complexities of *discantus*.[4] As noted in chapter 4, at Notre Dame in Paris, once the home of Leonin and Perotin and the fountainhead of polyphony, by 1408 *discantus* was no longer allowed to be inserted into chant regularly "with the exception of the boys, for the sake of training them."[5] Yet though polyphony was not regularly in use in the Cathedral of Rouen during the fourteenth century, the choristers of this cathedral sang a polyphonic motet to celebrate the Queen of England's visit in 1444.[6] Even the austere Augustinian canons of the Windesheim Congregation, who insisted on the purity of plainsong, allowed polyphony on Christmas Eve.[7]

The most vehement critic of polyphony during the fourteenth century was John Wyclif, the leader of the English Lollard reformist movement, who objected to almost every aspect of contemporary formal worship as unacceptable departures from the practices of the early church. In a treatise *Of feynid contemplatif lyf* (ca. 1380), Wyclif (or one of his followers) condemned "discant, organum, counternote, and small breaking [elaborate rhythmic subdivisions]" as "knacking" and "vain tricks … which stir vain men to dancing more than to mourning," which should be the only proper emotion evoked by the singing of sinful Christians: "But our carnal folk take more delight in such knacking and tattling in their bodily ears than in the hearing of God's Law, and the speaking of the bliss of heaven."[8] Wyclif also objected to elaborate plainchant, not just polyphony, both of which he thought obscured the sense of the words through musical artifice. Credited with the first translation of the Bible into English (in the 1380s), his musical views accordingly emphasized the intelligibility of the sacred texts, not luxurious indulgence of melodic or rhythmic elaboration, an objection that became a staple of antipolyphonic polemics in the following century. Wyclif argued that the Scriptures should be the center of Christianity, that papal claims to sovereignty lacked historical justification, that monasticism was corrupt beyond repair, and that the existing hierarchy ought to be replaced by "poor priests," bound by no vows, who would preach the authentic Gospel. After his death (1384), Wyclif was declared a heretic (1415), his body exhumed, burned, and scattered; the ensuing suppression of the Lollard movement limited the spread of his radical views, though he was later acclaimed as a forerunner of the Reformation.

A century later, Girolamo Savonarola in Florence raised many of the same musical objections as Wyclif. Like Thomas Aquinas a Dominican friar, Savonarola preached repentance, denouncing clerical corruption and the exploitation of the poor, setting in motion a puritanical movement of Christian renewal that ultimately had deep political repercussions throughout Italy (figure 5.1). At his instigation, the Florentines expelled their Medici rulers and installed a theocratic republic under his rule (1494–98). In pursuit of his vision of a purified church, he argued for the elimination of many clerical abuses, including the lucrative practice of votive masses whose endowments supported many clerics. His vision of purification also called for the expulsion of the Jews and new laws prohibiting gambling and sodomy.[9] He advocated the abolition of visual images and musical artifice that would undermine the purity of interior prayer: "And therefore the devil takes more care in this endeavor than any other; and therefore the demon has begun to turn aside mental prayer, which the elevated soul takes in contemplation, by introducing polyphony [*canti figurati*] and organs, which offer no delight except to the senses and of which there is no fruit."[10] He advised the Benedictine nuns "that this *canto figurato* was invented by Satan, that they should throw away those songbooks and organs."[11]

Figure 5.1
Girolamo Savonarola preaching in Florence, from his *Compendio di revelatione* (1496).

Conversely, when Savonarola visited Heaven in a vision, he described the music there as generally monophonic, listing a number of chants he heard sung by little children and nine choirs of angels; King David also sang for him and played the harp (presumably in a kind of simple homophony). The Virgin Mary conversed with him in Italian and promised that, though "for their iniquity they deserve every misfortune," through the prayers of the saints "the city of Florence will be more glorious and more powerful and more rich than ever," if it repent of its vanities.[12] Responding to his impassioned arguments, in March 1493 all polyphony was forbidden in the Cathedral and Baptistry of Florence. For Savonarola, polyphony was a vicious luxury of the ruling class: "Moreover, once in a while the tyrant keeps in the churches, not for the honor of God but his own pleasure, drunken singers (for they are plenty full of wine) who undertake to sing the Mass for Christ, and he pays them with the money of the commune."[13] These "drunken" singers were the expert performers needed for polyphony.

Savonarola did not take the view that polyphonic singing was merely "empty noise" (a view we shall shortly consider) but rather a seductive delight contrived by "the Devil, [who] under the guise of doing good, began to lead the monks into building beautiful churches, and conduct beautiful ceremonies, and give themselves to *canti figurati*; and all day to sing, sing, sing, so that nothing is left of the spirit; and thus the nuns all day with their organs, organs, organs, and there was nothing left; and in this way blight exterminated the greenery of prayers and of the spirit."[14] Savonarola thus had a high view of the power of music and therefore sought to use it for what he considered proper purposes. He taught his "boys" (*fanciulli*), the roaming bands of young enthusiasts for his cause, to sing simple songs, especially a rousing version of the Gregorian chant *Ecce quam bonum* ("Behold how good and pleasant it is for brethren to dwell together in unity"; ♪ sound example 5.1). This song in particular became the anthem of his movement, sung and even danced during the "bonfires of the vanities" he instituted during the Florentine Carnival, in which lutes and other musical instruments were burned alongside dice boards, lascivious pictures, playing cards, mirrors, dolls, and perfumes, in a great pyramid, "on the summit of which was the figure of Satan."[15] Besides this, Savonarola devised a number of other such rousing songs or *laude* whose use he encouraged.[16] Ironically, these were not strictly monophonic but were sung in a simple homophony, often replacing the ribald lyrics of the old Carnival music with a pious text (such as ♪ sound example 5.2). Evidently, by *canti figurati* Savonarola only meant complex polyphony, setting it apart from the simple homophony that was so familiar and popular that even he felt that it could be used innocently and positively.

Savonarola's antipolyphonic views were shared by his contemporaries in Florence, such as Fra Giovanni Caroli, who wrote in 1479 that in polyphonic singing

neither indeed can the words be adequately made out in that multiplicity of voices or sounds for the spirit to be greatly kindled in God, nor in that gaiety or swiftness of notes can its gravity any longer be preserved, but either the mind wanders away or at any rate dissolves into slumber. Therefore those polyphonies, which are both new and unheard-of and (if we will truly admit it) presumptuous, and so-called discants, lacking all harmony, do not much please.[17]

Though by the 1490s Caroli had become Savonarola's enemy and attacked his other ideas, they were of one mind about polyphony.

Similar struggles over polyphony took place in less famous places. In December 1486, the city elders of Görlitz ordered that all songs in its parish church be in plainchant "and not allow them to be turned into *hofereyen* [vanities] like they sing in the alehouses," meaning polyphony.[18] The parish priest, Johannes Behem, was so upset at this prohibition that he stormed out of the church, saying angrily to the elders: "You must not call them *hofereyen*! They are not *hofereyen*!" Behem had earlier clashed with the elders about the use of the organ in church, which they had likewise proscribed. He did not back down in this latest confrontation, which he first appealed to the local bishop, who ruled in favor of the city council "that one shall sing the chants of praise in the manner instituted by the Holy Fathers," meaning plainchant. The bishop did allow a "*carmen*" (presumably some kind of polyphonic song) to be sung on major feasts, but Behem was not satisfied and appealed the matter in person to the Holy See. Ultimately, the pope sustained Behem's contention that polyphony was allowable, but only over the strenuous objections of the city elders and bishop, indicating the strength of polyphonophobia. At several other places in Germany during the last decades of the fifteenth century, a considerable number of parents and educators expressed the view that "there is no need for singing in four parts, and for thus corrupting the tender minds of the young with empty noise," thereby wasting their time and energy and diverting them from more valuable studies.[19]

The famous Dutch humanist Desiderius Erasmus presented an even stronger version of this view. Writing in 1514, three years before the Lutheran crisis began, he reflected especially on his extensive experiences in England, where he was outraged by the "wanton whinnying and agile throats" of their elaborate polyphony. "In the Benedictine monasteries in England even youths, little boys, and professional singers are being maintained for this custom, who sing the early morning service for the Virgin Mother with the most elaborate vocal chatterings and with musical organs. It is choirs like these that bishops are expected to keep in the household. And, being all occupied with these things, they neither find time for literary studies, nor are they able to hear on what things true religion would depend."[20] As noted above, here too Erasmus's main objection was to the elaborate rhythms and melodies of the "whinnying" or "chattering," even more than to the multiplicity of voices, though he does single out "a depraved kind of singing which they call *fauburdum*," probably indicating some variety of the contemporary homophonic technique called *fauxbourdon*. Most of all, the net effect of polyphony is that "one or another [singer] mixing in with the rest, produces a stupendous bellowing sound, so that not a single word can be understood. ... It is noise alone that strikes the ears." Erasmus objected to the inordinate expense devoted to these practices, requiring great exactions from the people, who are "forced to listen to them, kept from the labors with which they feed their wives and children. ... Just calculate, I ask you, how many poor folk, barely clinging to life, could be supported with the stipends of singers?"[21]

Erasmus had to be careful to express these criticisms only after he was safely away from England, where any criticism of polyphony tended to be taken as evidence of being a Lollard sympathizer, hence a heretic. There, conformity with prevailing modes of worship, including polyphony, was a litmus test of orthodoxy. Indeed, a number of churchmen publicly replied to Erasmus's musical critique, engaging him in a lengthy controversy.[22] His musical opinions were shared to various degrees by the other protagonists of the Reformation. In 1522, Martin Luther abolished Gregorian chant and replaced it with congregational hymn singing; that same year, Huldreich Zwingli recommended the complete abolition of all music in the reformed churches in Switzerland.[23] Nevertheless, Luther defended polyphony; when the Elector of Saxony wanted to disband his chapel, Luther pleaded with him to keep it because "[this] art deserves to be maintained by princes and lords." He did not accept Erasmus's argument that polyphony is a futile luxury; instead, secular rulers, imitating King David, should "set singers also before the altar, that by their voices they might make sweet melody, and daily sing praises in their songs."[24] Luther admired Savonarola (whom Pope Paul IV less admiringly called an Italian Luther); one might compare the simple textures of Luther's hymns to those of Savonarolean *laude*, though these were often more dancelike and animated.[25]

As befitted the multiplicity of polyphonic art, the defenders of polyphony spoke in many voices in the later fifteenth century. They first took the offensive, arguing that those who condemned polyphony were fundamentally haters of music itself, intemperate zealots who betrayed their own lack of education and sensitivity. Here, Plato's authoritative defense of music became a common point of reference. In a letter of 1479, Marsilio Ficino (who first translated all of Plato's works into Latin) commented on the Platonic adage that "he is not harmonically constituted who does not delight in harmony," as cited by Augustine in his *De musica*. Ficino read this in terms of polyphony:

he who takes no pleasure in concordant sounds [*concentibus*, also signifying polyphony in general] in some way lacks concordant sounds within. I should say, if such a thing were permitted, that this man has not been put together by God, for God puts together all things according to number, weight, and measure. Moreover, I would say that such a man is no friend to God, for God rejoices in consonance to such an extent that he seems to have created the world especially for this reason, that all its individual parts should sing together to themselves and to the whole universe, and that the whole universe should resound as fully as it can with the intelligence and goodwill of its author.[26]

Accordingly, those incapable of musical delight tended to be dull-witted, lacking "penetration or judgment." Among many other statements of this view, Johann Walther gibed in 1538 that "the man whom music fails to thrill / is but a stick, inert and still / and worse yet than the savage beasts."[27] In this view, antipolyphonic diatribes were the worthless outpourings of vicious or stupid men.

Still, there remained another group who openly scorned music as unmanly. For instance, in Baldesar Castiglione's influential *Book of the Courtier* (first published in 1528), a dialogue about the qualities of an ideal courtier, signor Gasparo Pallavicino argued that "music, along with many other vanities, is indeed well suited to women, and perhaps also to others who have the appearance of men, but not to real men; for the latter ought not to render their minds effeminate and afraid of death."[28] In the ensuing dialogue, this frontal assault on music was immediately buried under "a great sea of praise for music," including stories of its prodigious powers recounted by Count Ludovico da Canossa: "Have you not read that music was among the first disciplines that the worthy old Chiron taught the boy Achilles, whom he reared from the age of nurse and cradle; and that such a wise preceptor wished the hands that were to shed so much Trojan blood to busy themselves often at playing the cithara? Where, then, is the soldier who would be ashamed to imitate Achilles, not to speak of many another famous commander that I could cite?" Speaking for the consensus of the participants, the Magnifico Giuliano de' Medici concluded that "music is not only an ornament but a necessity to the courtier."[29]

Roughly simultaneous with Castiglione's imaginary dialogue, the dispute over the manliness of music finally reached actual combat. After the music-loving Elector Palatine Frederick II heard Charles de Lannoy, Lord of Mangoval (and later Viceroy of Naples), assert that music was an effeminate trifle, unworthy of courtiers, Frederick appealed this dishonor of music to Archduke Charles of Austria (later Charles V).[30] Frederick was ready to "revenge the injury with swords," then and there, but the archduke judged "that the matter were better settled in a tournament." Accordingly, sometime during the mid-1510s in Brussels the armored protagonists, three on each side, fought on foot armed with "a lance with crown and a sword." The music lovers prevailed; "the false accusers of music were defeated at once, since there was no one would could endure the blows of the Palatine [Frederick]. For they retreated backwards, dodging every time he brandished his sword." Frederick was so infuriated that "he struck him on the temple with such fury that his eyes were blinded" and only the intervention of the archduke saved Mangoval from serious harm. "It was quite a sight to witness Mangoval with his companions, despoiled of his weapons, with black eyes and swollen cheeks and lips: he was sooner believed to be a monster than a human, and was laughed at by all."[31]

Still, to maintain their argument for the importance and powers of music, its defenders had to concede that those powers of music could be misused. Yet in the end they argued that the blame should fall on those who misused it, not on music itself. To the traditional praises of music these sophisticated defenders added a new standard that raised polyphonic art past the more easily apprehended beauties of simpler music. Thus, in his *Complexus effectuum musices* (written about 1474–75), the music theorist Johannes Tinctoris began by repeating earlier arguments: music softens hard hearts, lifts the mind above earthly things, banishes sadness, prepares for divine blessing, chases away the Devil, pleases God, and makes the church on Earth more like that in heaven.[32] But he added that the degree to which music makes human beings joyful depends on their discernment:

> For the more one has attained perfection in this art, the more one is delighted by it, since one apprehends its nature both inwardly and outwardly. Inwardly through the intellective power, through which one understands proper composition and performance, and outwardly through the auditive power, through which one perceives the sweetness of consonances. Only such are truly able to judge and take delight in music. … However, music brings less joy to those who perceive in it nothing more than sound, and who are indeed delighted only through the external sense.[33]

For Tinctoris, the proper apprehension of polyphony depends on *knowledge*, requiring that the hearer possess intellective power (*virtus intellectiva*) commensurate to the composer's intellect and art. Hence, the greater complexity of polyphony is justified by the way it increases joy through the use of intellect, adding a new intellectual element to Dante's praise of the paradisiacal aura of polyphony.

Other important figures continued to advocate polyphony during the controversies of the fourteenth and fifteenth centuries. Among them, the writings of Leonardo da Vinci opened new perspectives on these issues because he approached them as a practitioner of the visual arts as well as music. Then too, Leonardo came from the milieu of craftsmen rather than those privileged to have a "liberal education" and its higher studies in the quadrivium.[34] Because of its long-established place in this curriculum, music enjoyed a higher status than the visual arts, generally excluded from the liberal arts as merely "mechanical."

Though lacking this formal education, Leonardo was keenly aware of its ideals. His master Andrea del Verrocchio had been a noted "goldsmith, connoisseur of perspective, sculptor, engraver, painter, and musician"; in his turn, Leonardo became a skilled musician as well as an engineer and a supremely accomplished visual artist.[35] He had a beautiful singing voice and was a noted player on the lira da bracchio (figure 5.2), which he played while he sang improvised verses; according to Giorgio Vasari, "he was the best improviser of rhymes of his time."[36] This instrument, with its drone strings, was itself polyphonic, to which Leonardo added yet another voice, his own improvised singing. Indeed, it was in the capacity of a virtuoso lira player no less than engineer or painter that he was first called to Milan in 1494. There, he became friends with the music theorist Franchinus Gaffurius (who had been a student of Tinctoris) as well as the mathematician Luca Pacioli (whose works Leonardo illustrated).[37] Though conscious of his lack of conventional education and skill in ancient languages, Leonardo steeped himself in the liberal arts. He likely followed his friend Gaffurius's learned yet practical writings, such as his *Practica musicae* (*The Practice of Music*, 1496), which argued that "counterpoint may be adjudged to be most beautiful to the degree that the use for which it has blossomed forth is the more noble."[38]

During this period, Milan was a noted musical center where the Italian homophonic tradition mingled with the polyphonic art of Flemish composers; among them, Josquin des Prez was employed by the Sforzas in Milan from 1460 to 1470. Leonardo witnessed and participated in many musical events, especially the brilliant pageants, masques, festivities, and plays at the Sforza court. For instance, for the *Festa del Paradiso* (1490), Leonardo designed stage machinery that presented the twelve constellations of the Zodiac and the seven planets moving inside a gigantic hollow hemisphere, with "many sweet songs and sounds," probably including his own polyphonic performance as singer and player.[39] Leonardo also conceived and designed several innovative musical instruments, including a "viola organista" intended to "combine the polyphonic possibilities of the keyboard with the tone color of strings and thus would be something like an organ with string timbre instead of wind timbre."[40] He even designed a set of drums that could sound chords; his inventions, like his own performances, were conceived polyphonically.[41]

Leonardo's reflections on music come particularly in his notes toward a *Treatise on Painting* (*Trattato della pintura*), which (though unfinished) appeared after his death in a version put together by his pupil Francesco Melzi.[42] In the section he called "*Paragone*," he compared the relative status of music, painting, and poetry and argued that the conception of the liberal arts should be broadened to include painting: "Thus, because you have given a place to music among the liberal arts, you must place painting there too, or eject music; and if you point at vile men who practice painting, music too can be spoiled by those who do not understand it."[43] Here Leonardo joined a growing movement to broaden those arts considered liberal. Fifty years before, Leon Battista Alberti had argued that painting should be included among the liberal arts; in 1489, Antonio del Pollaiuolo

Figure 5.2
Marcantonio Raimondi, "Orpheus Charming the Animals" (ca. 1505). Ross Duffin (2015) speculated that in this image Orpheus, shown playing a lira da braccio, was a portrait of Leonardo da Vinci.

Figure 5.3
A detail from Antonio del Pollaiuolo's tomb for Sixtus IV (1489), showing Prospettiva (Perspective), a reclining figure surrounded by mathematical objects such as compasses, included alongside the traditional seven liberal arts (grammar, logic, rhetoric, arithmetic, geometry, music, astronomy).

sculpted Perspective among the seven traditional arts (figure 5.3).[44] Leonardo stood up for the "mechanical" arts, which he felt had been misunderstood and insufficiently valued. In the process of arguing that painting is in fact the elder sister of music and that both painting and music outrank poetry, he gave voice to a new understanding of music's polyphonic powers.

The heart of Leonardo's defense of painting is that it presents the attributes of its subject *simultaneously*, whereas poetry and music are forced to spread them out in time: "when the poet describes the beauty or ugliness of a body, he shows it to you part by part and at different [successive] times, while the painter lets you see it in one and the same moment." In this, the poet encounters the same difficulty "as would the musician, if he would sing by himself some music composed for four singers, by singing first the soprano part, then the tenor, and then following it by the contralto and finally the bass; from such a performance does not result the grace of harmony by proportions, which is confined to harmonic moments [chords]—this is precisely what the poet does to the likeness of a beautiful face when he describes it feature by feature."[45] Leonardo here describes not just an imaginary way of performing the four voices in sequence but reflects on the prevalent notational practice of his time. Though early manuscripts of organum (such as shown in figure 5.4) displayed the several voices arrayed vertically, it was not easy to check the intervals between voices because the parts were not precisely aligned spatially (as they are in a modern score). By about 1400, each voice was provided *separately* (figure 5.5), next to the others, often listed sequentially on the page. This change in practice doubtless had practical advantages, for it allowed each singer to read his part separate from all the rest. Often, these "part-books" would be arranged on a table so that those gathered around would see their own part conveniently before them. Yet, unlike modern notation, there was no "score" that would show all the voices together. Presumably, composers would compare the intervals sung by the various voices one by one to check that no unwanted dissonances would arise.[46]

Thus, when Leonardo described singing the voices one after another, he seems to have had in mind a part-book of his time and imagined what would result if one were to read it as if it were a normal text, singing each part sequentially, rather than simultaneously. His imaginary performance revealed that, precisely because of the simultaneity of several voices, music approaches the superior condition of painting and thereby exceeds the limitations of poetry. In this way, "many different voices, joined together in the same instant, produce a harmonious proportion that satisfies the sense of hearing to such a degree that the listeners remain stupefied with amazement, as if they were half dead. But still much greater is the effect of the angelic proportions of an angelic face represented in painting, for from these proportions rise a harmonic chord [*armonico concento*] that arrives at the eye in one and the same instant just as it does with the ear in music."[47]

Conversely, Leonardo seems to apply to music a concept of harmonious proportion that he draws from painting as it enumerates the various parts of a beautiful body. He was familiar with Alberti's *De re aedificatoria* (*On Architecture*, first published in 1485), which recommended that the laws of visual shapes (*figure*) be taken from music because "the same numbers that please the ears also fill the eyes and the soul with pleasure."[48] For his part, Leonardo argued that, as a painter gauges the relative proportions of a beautiful body, music "produces a body of many members whose whole beauty is contemplated by the listener [*contemplante*] in as many musical sections [*tempi armonici*] as are contained between birth and death and it is in these [successive] sections with which music enters the body of the contemplator."[49]

Here, Leonardo may have given one of the earliest spatialized images of musical form, made possible by the intimate relation he posits between the simultaneity of musical voices and the proportion of the successive sections they comprise.[50] The very notion of "musical form" (as it came to be called) depends on an extended analogy with visual arts that Leonardo began when he described music as "the figuration of invisible things [*cose invisibli*]" by comparison with painting as "the figuration of corporeal things [*cose corporee*]."[51] This remarkable, even paradoxical, yoking of the visible and the invisible represented Leonardo's attempt to endow painting with the ineffable aura of music; at the same time, he likewise endowed music with the solid sectional proportions that it drew from painting. To

Figure 5.4
The opening of Perotin's four-part organum *Viderunt omnes*, according to a manuscript version, whose score in modern notation is given in figure 3.4. Note that the bottom line shows the Gregorian chant (here only its first note), while the upper three lines show the organal voices, arranged vertically, though not as precisely aligned as a modern score.

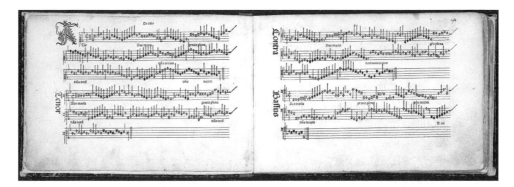

Figure 5.5
A part-book, from the *Harmonice musices odhecaton* (1501), showing four parts separately, not arranged vertically as in a modern score. Leonardo describes the different parts being sung sequentially, rather than simultaneously.

cement this novel comparison, he asserted that music, like painting (and also time itself), is based on *continuous* proportions—hence on geometry—thereby breaking with the ancient teachings that associated musical intervals (such as the octave, 2:1, or the perfect fifth, 3:2) primarily with *discontinuous* proportions between whole numbers, hence on arithmetic.[52] By downplaying the ancient connection between music and arithmetic, Leonardo showed how important he considered the priority of geometry in projecting his view of the sisterhood of music and painting.

By the time of the Council of Trent (1545–63), at which the Roman Catholic Church confronted the Reformation, the defenders of polyphony had largely prevailed, despite a long-standing myth that polyphony was on the verge of papal condemnation and was only "saved" after a performance of Palestrina's *Missa Papae Marcelli* (1562) so moved the conciliar fathers that they relented.[53] Though the issue of polyphony was raised in some draft documents, in the end the Council preferred to say as little as possible on the matter, except that church music should not be "lascivious or impure." Indeed, no less a polyphonophile than the Emperor Ferdinand I wrote the Council that "if the objective is that polyphony forthwith be removed from churches altogether, We are not going to approve it, for We consider that such a divine gift as music, which often kindles the souls of men—especially of those skilled or zealous in that art—to heightened devotion, ought in no way to be driven out of church."[54] Apparently, though, such high-level intervention was not necessary; at the Council, the advocates of polyphony seemed to prevail without much difficulty, leaving the regulation of music to local authorities.

The intellectual ecstasy of polyphonic art marked even the mystical experience of figures not generally identified with music. For instance, in his *Spiritual Exercises* (1522–24) Ignatius Loyola tended to emphasize visual (rather than auditory) cues in his guided meditations on the scenes of the life of Christ: those performing his exercises were encouraged to visualize those scenes in vivid detail in order to spur their inward response and personal engagement. Yet in his *Autobiography*, Ignatius described how, seeking "divine enlightenment" at Manresa during 1522–23,

his understanding began to be elevated so that he saw the Most Holy Trinity in the form of three keys [*teclas*]. This brought on so many tears and so much sobbing that he could not control himself. … He could not stop talking about the Most Holy Trinity, using many different comparisons and with great joy and consolation. As a result the impression of experiencing great devotion while praying to the Most Holy Trinity has remained with him throughout his life.[55]

Because the word "keys" (*teclas*) specifically refers to a keyboard instrument of his time, Ignatius seems to have visualized the Trinity as a triadic chord struck on a harpsichord or organ. But his description of why he wept emphasizes not so much what he *saw* as what he *heard*. His mystic audition shows the persistence and deep significance of the quest to *hear* God, three voices grasped as one.

II POLYPHONY TRIUMPHANT

WITHDRAWN
TOURO COLLEGE LIBRARY

6 *E pluribus unum*

Polyphonic compositions face the fundamental problem of connecting many voices into a felt semblance of unity, which composers accomplish in several ways. The technique of *cantus firmus* builds all the voices around a single melody, such as a Gregorian chant. *Canonic* technique allows all the voices to be repetitions of the same melody, perhaps slowed or speeded but essentially unchanged. Cantus firmus and canon can also be combined to build a further degree of unification, as we will see in a series of masterworks from the Middle Ages to the Renaissance.

Writing in 1475, Johannes Tinctoris defended polyphony as having a new dimension of intellectual content and inwardness that distinguished it from simpler kinds of music. But the heart of his defense was his enumeration of the new masters of polyphony:

> In our time we have experienced how very many musicians have been endowed with glory. For who does not know John Dunstable, Guillaume Dufay, Gilles Binchois, Johannes Ockeghem, Antoine Busnoys, Johannes Regis, Firminus Caron, Jacob Carlier, Robert Morton, Jacob Obrecht? Who does not accord them the highest praises, whose compositions, spread throughout the whole world, fill God's churches, kings' palaces, and private men's houses with the utmost sweetness?[1]

All these have attained "immortal fame," which Tinctoris (quoting Virgil) associates with "the task of virtue, to prolong fame by deeds." For Tinctoris, these composers' sweetness and glory is, more than anything else, the mark of their achievements and significance. Implicitly, "by their fruits ye shall know them": the masterworks of this international (though notably Flemish) constellation of composers were the paramount justification of their polyphonic art.[2]

What, then, were the achievements that deserved such glory? I will illustrate them by considering three works spanning the fourteenth through sixteenth centuries. Each addressed in somewhat different ways the fundamental problem of polyphony: how can several independent voices become a unity? To some extent, all rely on one fundamental technique: a single melodic line (often a Gregorian chant)—the cantus firmus—forms the basis out of which all the other voices are built and with which they must concord. The unity and ancient authority of the cantus firmus thus unifies all the voices built from and around it. Already the earliest forms of organum mentioned above use this technique: the cantus firmus is heard in long notes and the various organal voices form simple consonances with it on the beginnings of each measure, though they are allowed to make passing dissonances as they move from measure to measure. This technique is especially clear in Perotin's *Viderunt omnes* discussed earlier (figure 3.4), in which the cantus firmus is slowed almost to a standstill, sustaining a single note over many measures of a dance-like rhythmic pattern in the other three voices.

Guillaume de Machaut used a variant of this unifying cantus firmus technique in his *Messe de Nostre Dame* (*Mass of Our Lady,* written about 1360), the earliest extant setting of the common parts of the Catholic Mass (those generally required on all occasions) written by a named composer.[3] Machaut was both a poet and composer of the highest importance, perhaps the last person to combine those different excellences to such degree. A canon of the newly built cathedral dedicated to Notre Dame in Reims, his mass paid special homage to the Virgin Mary, the object of an ever-growing cult from the twelfth century onward. There is evidence that Machaut intended this to be a special votive mass to be sung at memorial services for his brother and himself at Reims, conceived as a musical monument that would (like the perpetual services it would accompany) form a special repository of his artistic and spiritual legacy. For centuries thereafter, though, it lay mostly unread in manuscript form, unknown to many generations of subsequent composers (Tinctoris, for instance, does not mention Machaut).[4]

Despite its early date, Machaut's mass sounds strikingly "modern," a quality we noted of Perotin's organum. Machaut's mass is austere and dissonant, so complex and rebarbative that it seems closer in style to Igor Stravinsky's *Mass* (1958) than to many other masses written during the intervening centuries. This quality emerges from the very beginning of the work, his Kyrie I (figure 6.1; ♪ sound example 6.1). Where Perotin offered one clear voice (intoning the chant) surmounted by three interweaving voices, Machaut's four voices proceed en bloc, none clearly standing out above the rest, though the ear may most easily notice the highest sounding voice.

Compared to the clear separation between organal voices and the cantus firmus in Perotin's organum, the cantus firmus of Machaut's Kyrie I is buried within its texture in the tenor (here shown in the bottom line, though its pitches actually lie higher than those of the line above it, the contratenor); as *tenere* means "to hold," here the tenor holds together the contrapuntal fabric. Machaut's tenor melody is a well-known Gregorian chant, which he transformed from unmeasured chant notation to 3/2 meter (in terms of modern time signatures; see figure 6.2). According to contemporary rhythmic theories, meters based on three beats (the "most perfect number because it takes its name from the Holy Trinity, which is true and pure perfection") were considered "perfect," compared to those based on two beats considered "imperfect."[5] Thus, 3/2 may represent a synthesis between perfection and imperfection, as the person of Christ united divinity and humanity.

Figure 6.1
Guillaume de Machaut, Kyrie I from *Messe de Nostre Dame* (♪ sound example 6.1). Text: "Lord, have mercy."
Note the alignments on D–A marked with boxes; the isorhythms are detailed in figure 6.2.

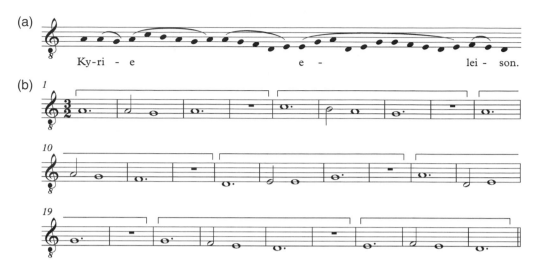

Figure 6.2

(a) The Gregorian chant *Kyrie cunctipotens genitor* transcribed in unmeasured notation (♪ sound example 6.2a); (b) the tenor voice in Machaut's Kyrie I, which contains the same twenty-eight pitches in the same order, but now presented in 3/2 meter (♪ sound example 6.2b). Note the "isorhythmic pattern" or *talea* repeated every four notes, indicated by brackets: the same sequence of rhythms in that voice stated in measures 1–4 is then repeated in measures 5–8, 9–12, and so on. In contrast, the contratenor (the voice shown just above the tenor in fig. 6.1, though singing generally lower pitches) has an talea three times as long: the same pattern of rhythms in measures 1–12 then repeats (with some variations) in measure 13–24.

Alongside the melodic underpinning of the cantus firmus in the tenor, Machaut used the contemporary technique of *isorhythm*, shown in figure 6.2: a recurring rhythmic pattern (called a *talea*, in this case consisting of four notes) that cycles seven times through the twenty-eight notes in the tenor part of Kyrie I. Though the contratenor part does not use a chant melody, it too has its own isorhythmic pattern three times as long as the tenor's. The top two voices (the triplum and motetus) do not have isorhythms but are more freely written, adapting the general procedure that already applied in Perotin: all four voices should sound consonances with the tenor at the beginning of each measure.

Because the cantus firmus tenor circles around the notes D and A, these two pitches have a special importance for Machaut's Kyrie I. For instance, in the first bar all four voices respectively sound either the notes D or A, to which they return in several subsequent measures (marked in figure 6.1). Between these striking moments of alignment, the other voices often move with daring freedom against the tenor, forming dissonant intervals in passing, sometimes "hiccupping," the literal meaning of the term "hocket" (the rapid exchange of notes between voices, one resting while the other sounds, as in measures 10 and 22), though always returning to consonance at the beginning of the next measure.

The recurrent consonances (especially the D–A concord) might be compared to planets periodically returning to their initial configuration. This comparison could well have appealed to Machaut, who studied astronomy along with music in the course of his liberal education in the quadrivium. Ptolemaic astronomy had developed an intricate and accurate system for recording and predicting planetary positions, especially the conjunctions of planets (points at which they appear to lie at the same position in the celestial sphere), their oppositions (appearing 180° apart) and quadratures (90° apart), as illustrated in figure 6.3.[6] These striking alignments between planets are analogous to those between Machaut's voices; both are recurrent phenomena whose timing reflects the larger cosmic or musical rhythms at work. Though astronomical conjunctions tend to recur in simple multiples of a basic unit of time, these musical conjunctions recur with less predictable accuracy, guided by the return of the chant itself to D or A.

These musical recurrences have the force of cadences—that is, punctuations or closures—to greater or lesser degrees: the final resolution on D–A has a greater degree of "arrival" than the earlier moments in the movement in which those notes are passed through without stopping (marked in figure 6.1). As such, those transient cadential moments give a varying but perceptible punctuation to what might otherwise seem an unstructured texture. But the isorhythmic patterns are scarcely audible unless one is following each voice closely and even counting its rhythms. Later in his mass, Machaut chooses even longer and more complex isorhythmic patterns, whose effect likewise is not immediately audible (see figure 6.4; ♪ sound example 6.3).[7]

These patterns, though deliberately chosen, may not have been intended primarily to make a conscious impression. Arguably, they do affect the listener through their successive variations, if only by contributing to a sense of accumulating and subtle variation that heightens the effect of repetition. As with the novelties of Notre Dame polyphony, here again the closest analogy may be to architecture. Many aspects of the Gothic style involve structures not designed to be visible, such as the hidden supports that allow those cathedrals their extraordinary height and luminosity, not to speak of the visibly "flying" buttresses. Such invisible structures are architecturally necessary, though discernible only to those knowledgeable in the craft. Other less obviously necessitated structural and ornamental devices may have been intended for the eye of God, or perhaps for those whose training and aptitudes prepared them to appreciate their ingenuity. Consider, for example, the gargoyles, many of which are so high up or placed so remotely in these Gothic cathedrals that they cannot really be seen except by those who have access to the higher stories of the church, such as the cathedral clergy (like Machaut) or perhaps only the builders themselves. Yet those gargoyles are intricately (often humorously) carved, despite their invisibility, signaling the importance accorded to what might otherwise seem imperceptible details. Perhaps this would not prevent them from also affecting most of the faithful in the church, for whom the cathedral was a "Bible for the illiterate," making manifest in space and image the truths and stories they could not read. Even someone who could not discern isorhythmic patterns or architectural proportions in any explicit way might be deeply affected by them below the threshold of awareness. Thus, these esoteric and

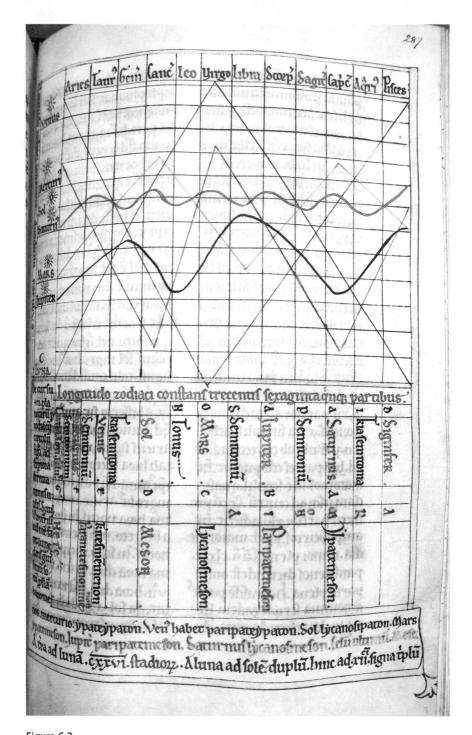

Figure 6.3
A twelfth-century manuscript of Bede's *De temporibus* showing the motion of the planets against the zodiac. By permission of the Master and Fellows of St. John's College, Cambridge (MS I.15 f. 145r).

Figure 6.4

Machaut, Christe eleison from his *Messe de Nostre Dame* (♪ sound example 6.3). Text: "Christ, have mercy."
Here, the tenor's isorhythmic *talea* pattern contains eight notes, this most "imperfect" of numbers ($2 \times 2 \times 2$)
underlining the plea to the Savior of human imperfection, as if extending the four-note *talea* used by the tenor
in Kyrie I.

Figure 6.4 (continued)

exoteric aspects of music and architecture would all have their place in the total experience of the work.

Indeed, Machaut does not always rely on arcane isorhythmic patterns, especially in the parts of the mass that address the people most directly, particularly the Gloria (beginning with the message "glory to God in the highest" sung by the angels announcing the birth of Christ to the shepherds) and the Credo (the setting of the Nicene Creed that summarizes the doctrinal precepts of Christian faith).[8] For these sections, Machaut chose a rather homophonic, rhythmically unified declamation that gives a general coherence to the Gloria and Credo while avoiding any strict patterns.[9] In these sections, he follows the text closely, as is appropriate for these doctrinal statements whose extended texts call for clear articulation. For instance, responding to the priest's chanted *Gloria in excelsis Deo*, the voices take up the text "and peace on earth" (*et in terra pax*) (figure 6.5; ♩ sound example 6.4), which Machaut sets with direct simplicity.

Though the voices move in contrary directions, the unanimity of their texture is paramount, what later would be called "block chords" moving together toward a characteristic cadence (measures 4–5) that gives musical punctuation to their utterance. Machaut applies the same treatment even more starkly to the words "Jesu Christe" in the Gloria, the first point in the mass that explicitly names the Savior (figure 6.6; ♩ sound example 6.5). This name is set with utter simplicity in long, equal chords that bring it into the highest relief against the strongly rhythmic declamation applied to the rest of the text.

This may be the first instance of "word painting" in our examples: the name of Jesus is set off rhythmically, harmonically, and texturally to underline the text's assertion that "you alone are most high." Machaut reserves his only other use of this stark declamation for the words in the Nicene Creed avowing that Jesus was born "from Mary the Virgin" (figure 6.7; ♩ sound example 6.6). By according these words the same treatment as the name of Jesus, Machaut clearly offers comparable honor to his mother, as befits a cathedral (and

Figure 6.5
Machaut, *Messe de Nostre Dame*, Gloria, measure 1–5 (♪ sound example 6.4). The "double leading-tone cadence" in measures 4–5 involves two voices moving by semitones to their final notes (from C# to D or G# to A), while the third (the contratenor) moves downward by step from E to D.

Fig. 6.6
Machaut's setting of the words "Jesu Christe" in the Gloria of his *Messe de Nostre Dame* (measures 90–97) (♪ sound example 6.5); note the double leading tone cadence in measures 96–97. Text: "You alone are most high, Jesus Christ."

mass) honoring her. Thus, Machaut's polyphony includes the most independent, dissonant writing as well as simple homophony. He treats the unifying factors of cantus firmus and isorhythm as hidden architectonic factors undergirding the edifice of his mass. His hieratic, often dissonant design frames the names of Jesus and Mary, connected in the mystery of incarnate personhood.

By comparison to Machaut's mass, Johannes Ockeghem's *Missa prolationum* (written a century later, in the second half of the fifteenth century) has a more consonant, even sweet sound. One of the famous composers on Tinctoris's list, Ockeghem did not use preexisting Gregorian melodies to provide the cantus firmus in this work. He probably wrote his own melodies and used them to create an extraordinary contrapuntal creation by making the whole mass out of what came to be called *mensuration canons*. As mentioned above, by the mid-thirteenth century canons appeared in secular music such as "Sumer Is Icumen In," a single theme chasing itself through several voices (figure 3.5; ♪ sound example 3.4). The technique of canon unifies several voices even more than cantus firmus: in a canon, the identical intervals are shared by *all* the voices, whereas in a cantus firmus composition, each voice may sing different intervals, though regulated so as to remain consonant with the cantus firmus at the beginning of a measure. Because of these compositional advantages, canon became widely used in sacred music, in serious contexts far removed from the playfulness of rounds.

In a mensuration canon, different voices repeat the same theme but at various rates of speed (different time signatures called "prolations"), so that the same melody will last a different duration for each voice. For example, in Ockeghem's Kyrie I (figure 6.8; ♪ sound example 6.7), the four voices are conceived as two simultaneous duets, in each of which one voice sings the same melody as another, but at a different speed. Essentially, the same melody plays against itself but in two different meters: the top voice in 2/2, the second in 3/2 (forming the first pair); the third in 6/4, the fourth in 9/4 (using modern terminology). Even though all four voices begin at the same time singing the same two melodies, because of the different meters they take different lengths of time to complete the melody. Yet the underlying rhythmic pulse is the same: the half note that is shared by all.

Despite the amazing artifice of this construction, most listeners would not notice how it was done or even that it was going on. To make it work, Ockeghem had to choose melodies that would avoid excessive dissonance when overlaid against themselves in canon. To achieve this kind of unity between the polyphonic voices, he thus was under pressure to make the melodies (and hence their combinations) rather consonant. By comparison, Machaut's voices are freer and hence more often dissonant during the course of each measure. For future reference, we record the consequence: the greater the degree of unity of the composition, the more carefully it aims for consonance and the more strictly it controls dissonance. Then too, Machaut's voices proceeded in a clearer temporal synchronization because the isorhythmic units on which he built were coordinated within a single larger time frame or meter. A hearer capable of discerning Ockeghem's mensuration canons would need to keep track of several rates of musical "time" at once, not just different isorhythmic

Figure 6.7
The words *ex Maria Virgine* ("from Mary the Virgin") in the Credo of Machaut's *Messe de Nostre Dame* (measure 68–74) (♪ sound example 6.6); note the cadence in measures 72–73.

Figure 6.8
Johannes Ockeghem, Kyrie I from *Missa prolationum* (♪ sound example 6.7). The top two voices share the same melody; the bottom two share a different melody. Each voice proceeds in a different meter (2/2, 3/2, 6/4, 9/4) while always sharing the same underlying pulse (the half note).

patterns. Again one thinks of the composer addressing a more-than-human mind capable of "hearing" at four different time-rates at once. If so, not only would that mind know many things at once (as Thomas Aquinas held of God and the angels) but many different *kinds* of time unfolding the same melody at different rates, able to "telescope" those different meters into the original melody from which they all came.

Ockeghem's *Missa prolationum* was a unique tour de force; other composers used mensuration canons in less extended contexts, but not for an entire mass. Still, the basic technique of imitation or canon became very widely used in the succeeding century; many masses by masters such as Josquin des Prez were largely composed of canons based on a theme from Gregorian chant or sometimes even on secular songs. Generally, these canonic settings did not involve different mensurations, such as Ockeghem used, but more transparently retained the rhythm and meter of the original melody. Occasionally, for special effect a canonic melody could be played against itself speeded up by a factor of two or even four (later called "diminution") or slowed by the same factor ("augmentation"). This simpler subset of mensurational possibilities allowed a further dimension of contrapuntal writing that could underline special moments in a composition, devices to which we shall return later.

The possibilities of simpler and more straightforward forms of imitation allowed the listener to recognize this device more easily than the more recondite artifices of isorhythm or mensuration canons. Josquin, who was said to have studied with Ockeghem, used these more recognizable forms of imitation and canon to create a polyphonic texture intensely admired by his contemporaries. Sometimes he used a cantus firmus that was even simpler than a chant; his *Missa La sol fa re mi* (first published in 1502 but written sometime between 1470 and the 1490s) is built on just that sequence of pitches, almost a pure descending scale from *la* to *re* except for the reversal of the final two notes (see figure 6.9a; ♪ sound example 6.8). This was one of the earliest examples of a technique Josquin pioneered that came to be called *soggetto cavato*, a melody literally "carved" (*cavato*) from the syllables of some words. Thus, this carving allows a kind of musical cryptogram. In this case, the words have come down with two alternative meanings: in old French ***Laise faire moy*** (in Italian ***Lascia fare mi***), literally "leave it to me," either the words of someone confidently promising to pay what he had promised; alternatively, someone trying to put off those pestering requests ("just leave it to me," but not really intending to do anything). The true story remains unclear; commentators have advanced various possibilities.[10] Perhaps Josquin's patron Cardinal Ascanio Sforza, strapped for cash, either promised with these words to pay his protégé or (alternatively) tried to weasel out of his debts. In another version, the Turkish Prince Cem Sultan, held by the pope in luxurious hostage at Rome, used these ambiguous words to promise to overthrow his half-brother Bayezid II (then ruling the Ottoman Empire) or alternatively to feign fulfilling those demands. To add further levels of mystification to the story, Cem persistently refused to convert to Christianity, and Josquin

(a)

la   sol   fa   re   mi

(b)

Figure 6.9
(a) The subject of Josquin des Prez's *Missa La sol fa re mi* (♪ sound example 6.8); a pure descending scale would have ended *mi re*. (b) Josquin wearing a turban.

was portrayed wearing a turban (figure 6.9b); was this merely a contemporary fashion, or was the composer mocking the prince?

Whatever the true words or story (whose ambiguity he may have relished), Josquin here crafted an arrestingly simple melody, easy to grasp and remember, which he then used conspicuously throughout the whole of this mass. In many places, the subject floats prominently in long notes so clear that an untrained ear could recognize it; for instance, at the very beginning of his Kyrie I, the subject is passed clearly from the superius to tenor to bassus, setting the words "Lord, have mercy" (figure 6.10; ♪ sound example 6.9).

Josquin does not stop here; he sets almost every line of the mass to some form of this same subject, which he applies to the words "peace on Earth to men of good will" and "I believe in one God, the Father almighty" (figures 6.11–12; ♪ sound examples 6.10–11) no less than "Lord, have mercy" (figure 6.10; ♪ sound example 6.9). As the work unfolds, Josquin uses this subject to set the whole gamut of theology, from "he was crucified for us," to "he rose again," "hosanna in the highest," and "grant us peace." In so doing, Josquin shows that his interests transcend telegraphing or echoing the meaning of the text in any obvious musical way. Instead, his unitive musical project invites the listener to hear the whole spectrum from confession and prayer to glorification as emanating from one single motive.

Figure 6.10
Josquin, Kyrie I from *Missa La sol fa re mi* (♪ sound example 6.9). Note that the subject appears in its original form in the superius (measures 1–5) and in the bassus (10–13), then transposed a fifth higher (starting on E) in the tenor (5–8).

Figure 6.11

Josquin, opening of the Gloria from *Missa La sol fa re mi* (♪ sound example 6.10). The subject (transposed to begin on E) runs through the tenor. Note the subject sung forward, then backward (retrograde) in the altus, measures 7–8, making a palindrome. Text: "And peace on Earth to men of good will."

Figure 6.12
Josquin, the beginning of the Credo from *Missa La sol fa re mi* (♪ sound example 6.11). The tenor sings the subject (transposed to begin on E) in greatly slowed form ("augmentation"). Text: "[I believe in one God], Father Almighty, maker of heaven and Earth."

As he does so, he sometimes will play this subject in different rhythms or immensely slowed down in one voice, such as at the beginning of the Nicene Creed (figure 6.12; ♪ sound example 6.11). He thus uses some of the mensurational devices Ockeghem employed in his *Missa prolationum* but in a way more readily grasped by the listener, who here could recognize the immense structure of belief articulated in the Creed emanating from this greatly prolonged subject, as if time itself could be comparably slowed to allow us to contemplate it from a divine perspective.

Nor is this the limit of Josquin's contrapuntal artifice; at times, he will use the subject (in one or another of its variant forms) sung backward, "retrograde": *mi re fa sol la*. When this is preceded by the subject in its forward reading (*la sol fa re mi*), a palindrome results, a sequence of notes the same whether forward or backward, another image of eternity. Compared to the open presentations of the subject, these retrograde or palindromic passages are hidden (such as the one noted in figure 6.11), adding to the outward or exoteric beauty of the work another level of esoteric art, intended for close students of the work or for God himself as supreme artist, artificer of "all things visible and invisible." But these more hidden beauties are subsidiary to Josquin's larger project, which seems to depend on the near-constant sounding (and audible recognition) of the subject in its simplest, most direct form: it appears over two hundred times. One is reminded of an endlessly repeated peal of bells, ringing over and over the same sequence of notes, yet surrounded by an ever-changing garland of counterpoint so that one hears the work's unity unfolding in the midst of ceaseless variation.

As princely patrons sought new ways to project their power, they commissioned polyphonic works with large numbers of voices. The Medici in Florence, seeking the status of
grand dukes, mounted a "charm offensive" involving extravagant polyphonic works for
many voices, which succeeded so well that they elicited a rival work that defended English
polyphonic pride. As sumptuous as these works were, their large number of voices necessitated a certain simplification of their harmonies: to avoid excessive dissonance, the voices
have to share notes. As polyphony reached these limits, it paradoxically generated newly
unified textures out of the interplay of many voices.

Well before Savonarola, elaborate polyphony was already an important part of the self-
projection of the Medici, whose renowned court composer, Heinrich Isaac, gave luster to
their ceremonies and celebrations. After the execution of Savonarola, the Medici returned
and resumed their project of aggrandizing their city and family through commissioning
ever more complex polyphonic music. In the sixteenth century, these efforts reached a
peak in the work of Alessandro Striggio, who around 1565 composed a motet, *Ecce beatam
lucem*, for forty independent voices. He also composed a forty-voice *Missa sopra Ecce si
beato giorno* that included an *Agnus Dei* for no fewer than sixty independent voices. These
extraordinary works were part of a Medici musical-political offensive: in late 1566, they
sent Striggio on a midwinter journey through the Alps (accompanied by a pack donkey) to
Vienna to present this mass to the Holy Roman Emperor Maximilian II as part of their efforts to gain a new royal title as grand dukes of Tuscany, which they duly received in 1569.
Striggio's journey also included visits to royal courts in Munich, France, and London,
where his new works were also performed.[1]

Thus, the Medici judged these new extremes of polyphonic art to be the ideal medium
to project their grandeur by implying that their ability to command an unprecedented number of independent musical voices paralleled their wealth, political might, and dynastic
worth. Ironically, at that point the Medici were strapped for cash, but they had found an
ingeniously cheap stratagem that only required sending a single man with a score, capable
of organizing the needed singers when he arrived. Indeed, Striggio's forty-voice motet
Ecce beatam lucem gives an overwhelming impression of magnificence. Its voices are

organized into five choirs, each comprising eight voices, which could be separated spatially to heighten the polyphonic impression. Thus, one scarcely hears any single voice by itself but only as part of a much larger texture, often involving two or more of the choirs. For instance, the opening statement by choir 1 is answered by two voices from choir 4 and the whole of choir 5; in response, choirs 2 and 3 then enter en masse (figure 7.1; ♪ sound example 7.1). These entries would have seemed to come from different spatial directions, as well, using the new dimensionality that Venetian works for several antiphonal choruses had already added to music.[2]

Because of the sheer number of separate voices, the resulting musical texture seems rich and gorgeous, just as the Medici hoped; the hearer's inability to distinguish the lines clearly (compared to four voices, say) gives rise to a sumptuous impression of iridescent waves of sound. Ironically, though, this seemingly rich texture turns out to be harmonically very simple. The issue is this: the sheer number of voices could lead to excessive dissonance between them unless an almost completely consonant harmony were imposed. By this time, composers increasingly relied on the harmony of triads, so that at any given moment only three separate pitches could be sounded (apart from brief passing notes in various voices, bridging adjacent consonances).

Thus, if forty voices were sounding at once, at any one moment basically they could only sing three different notes between them, leading to great redundancy between these "independent" lines. Striggio artfully masks this limitation by spreading out these notes in every possible register available through the number of voices; if twelve or so voices are singing a certain pitch, they might do so in four different octaves, three voices each, mitigating the sense that they are merely doubling each other. To support their pitches, Striggio added a forty-first voice, a bass line called a *continuo* (also called "thoroughbass" or "figured bass") to be played by various instruments (such as an organ or harpsichord), reinforcing and summarizing the harmony of the whole. The extreme simplicity of this line (at the bottom of figure 7.1) underlines the paradox: the more the number of consonant voices, the simpler the harmony they can produce. This simplicity marks the climactic moment when all five choirs and forty voices enter simultaneously (figure 7.2; ♪ sound example 7.2): though the motion in various voices gives textural feeling, the overall impression is of unanimous declamation.

When Striggio presented his novel works at Queen Elizabeth I's court in 1567, they excited not only wonder but the desire to emulate or even excel them. One of her dukes, "bearinge a great love to Musicke asked whether none of our Englishmen could sett as a good a songe."[3] Among the English "virtuosos in the profession of music," Thomas Tallis, "beinge very skilfull was felt to try whether he could undertake ye Matter." His celebrated forty-voice motet *Spem in alium* "so farre surpassed ye other [Striggio's]" that the duke took a gold chain from his own neck and gave it to Tallis, rewarding English pride through its feat of contrapuntal art. His countermotet not only shows that others could match Striggio's feat but pushed polyphony to other extreme possibilities.

Figure 7.1

The opening of Alessandro Striggio's motet *Ecce beatam lucem*, whose forty voices are organized into five choirs of eight voices (♪ sound example 7.1). The opening statement by choir 1 is answered by choirs 4 and 5 (measures 2–5), then by choirs 2 and 3 (measure 5). Note the additional (forty-first) voice, a continuo that underlines the harmony of the whole. Text: "Behold the blessed light."

Figure 7.2
The climactic entry of all forty voices simultaneously in Striggio's motet *Ecce beatam lucem* (measure 107)
(♪ sound example 7.2). Text: "Draw us from here straight to Paradise."

Though Striggio had restricted his voices to enter in large blocks or by whole choruses, Tallis exploited a far larger spectrum of polyphonic possibilities. He organized his forty voices into eight choirs of five voices each (compared to Striggio's five octets of voices) and went beyond Striggio's use of the choirs in blocks. Tallis's beginning is particularly magical: he has the forty voices enter one by one, starting with the highest voices in choir 1, then slowly and successively including the rest. The effect suggests light gradually emerging out of darkness, as the sounds come forward in the silence. Until the fourth measure, only a lone soprano and alto sing, gradually joined by more and more until twenty voices are singing together (measure 24). Then the succeeding voices enter in turn as the initial twenty gradually fall silent. Though Tallis's singers would not have had it in front of them, a score shows the wave of voices slowly moving downward (figure 7.3; ♪ sound example 7.3). If, as seems likely, these voices were arrayed in space left to right, the hearers would have had a spatial experience of the sound gradually moving around them in that same directional way.

After this gradual unveiling of the forty voices, Tallis also avails himself of block entrances of several choirs at once, as had Striggio, using the force of their simultaneous entries to create a dramatically new texture. Spatially, these responsive entries of whole choirs (or even groups of them) allow a different kind of stereophonic interplay than the subtle rotation through the voices at the beginning, which moved from left to right. Later in the motet, Tallis creates a counterrotation that begins in the fortieth voice and moves upward (and leftward) through successive entries by the voices (beginning in measure 35). Using this kind of gradual buildup of polyphonic complexity, Tallis several times builds to a texture in which all forty voices sound at once. At several points, but most notably in the final pages, Tallis creates a rest in all forty voices—whose silence itself seems polyphonic—answered by the simultaneous entrance of all the voices (figure 7.4, measures 121–122; ♪ sound example 7.4).

Though Tallis left no indications of the dynamics he expected, at this point the text "look upon our lowliness" suggests that the entrance of all the voices should be soft, as would befit a moment of humble supplication. Compared with Striggio's use of a text about light and glory, Tallis consciously set his forty voices to express humble adoration and hope. The subtext seems to be that Tallis's motet exceeds Striggio's not only in polyphonic subtlety but in true religiosity, in proportion as pious humility is worthier than self-projected grandiosity and (implicitly) as Protestant England stands above Catholic Italy. After the solemn silence in measure 121, Tallis's final pages interweave all forty voices in a texture that has more inner motion in the individual voices than Striggio chose to use. To be sure, both composers faced the same stringent demands for consonance, limiting the possible pitches the voices could sing. Yet where Striggio chose a harmonically simplified declamation that tied the voices more closely together, Tallis allowed greater freedom of motion in the individual voices in order to create a radiant harmonic stasis, as if the interplay of the Many revealed the One behind them all.

Spem in alium nunquam habui

Thomas Tallis
(*c.*1505–1585)

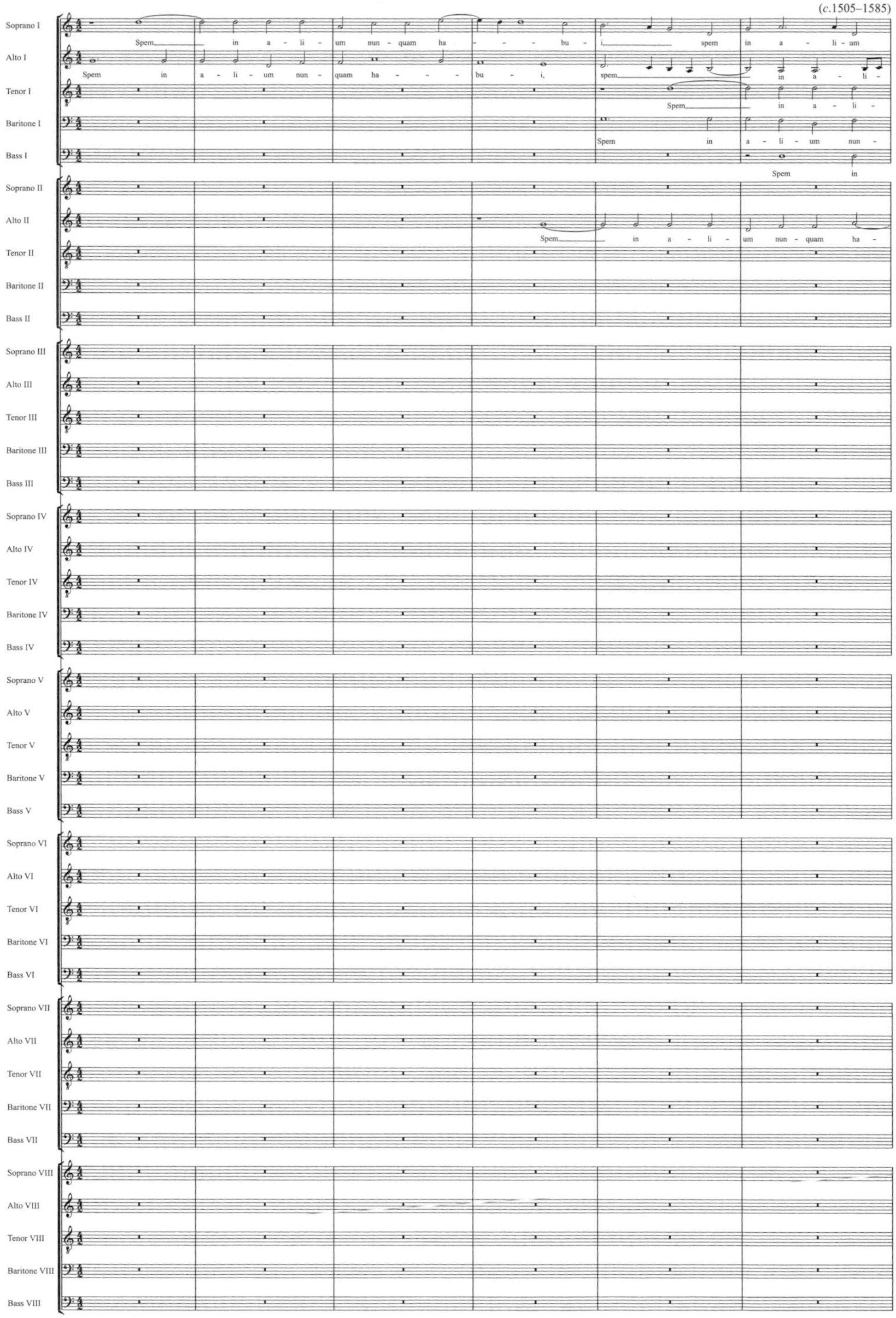

Figure 7.3
The opening eighteen measures of Thomas Tallis's motet *Spem in alium*, showing the gradual entry of the first twenty voices (♪ sound example 7.3). Text: "[I have] no other hope."

Figure 7.3 (continued)

Figure 7.3 (continued)

Figure 7.3 (continued)

Figure 7.4
The simultaneous entry of all forty voices, after a general pause (measures 121–122), from Tallis's *Spem in alium* (♪ sound example 7.4). Text: "Look upon our lowliness."

Besides its many performances and recordings over the centuries—limited by the requirement of finding forty skilled singers capable of taking on the work—Tallis's motet entered into a new phase of its posthumous life in 2001, when the Canadian installation artist Janet Cardiff created a new version called *Forty-Part Motet*. Having miked each singer individually during a recording of *Spem in alium*, her installation plays these forty tracks back through individual speakers, each placed at head height, all arranged in a large oval (figure 7.5). The sound track for each voice is a loop that also includes three minutes of random conversation between the singers before and after they sing the motet; entering the installation at that point, one would hear these fragments of speech coming from the various speakers. After a silence, the motet itself begins. One can wander between the speakers, sometimes hearing them all from a central vantage point, sometimes coming up close to a speaker and hearing its separate voice close up. The individual recordings were acoustically "dry" so that the work takes its overall resonance from the space in which it is installed, its acoustic ambience and reverberation or lack thereof.

Cardiff herself first heard the motet in a CD recording: "Immediately I was infatuated with the music but at the same time was totally frustrated because I wanted to hear the forty different harmonies, I wanted to hear how they would go up and down and all around and I just envisioned right away this choir around me and I know that some choirs actually do this singing in the round and that was one reason that I wanted to place it in an oval."[4] Her installation opened the complementary possibilities that one could hear the work "in the round," its many voices cohering, yet still allowing the listeners to walk up to one of

Figure 7.5
The installation of Janet Cardiff's *Forty-Part Motet* (2001) presented by the San Francisco Museum of Modern Art at Fort Mason (2015–16). Photo credit: jka.photo.

the speakers and attend to it apart. "I want the audience to be able to experience a piece of music from the viewpoint of the singers. Every performer hears a unique mix of the piece of music. Enabling the audience to move throughout the space allows them to be intimately connected with the voices. It also reveals the piece of music as a changing construct."[5] Elsewhere, she noted that "most people experience this piece now in their living rooms in front of only two speakers. Even in a live concert the audience is separated from the individual voices. Only the performers are able to hear the person standing next to them singing a different harmony. I wanted to be able to climb inside the music."[6] Though these voices now emerge from speakers, not living humans, one can encounter the sound of one voice with strange new intimacy: "It feels like it's on the back of your neck when you're listening to it. There's a sense of safety in technology, but at the same time there's a sense of connection. You don't lose the connection of the human voice, but you get the safety of not having to stand right next to that person."[7]

Cardiff's installation provoked strong reactions. In New York, press reports described people emerging "wobbling, blissed out, a few in tears … 'too raw,' said another young woman … 'transcendent.'" Another woman noted that "each speaker is a different person. It's not something you think about: you feel it."[8] Here, the personhood of the voices—discussed earlier in medieval philosophical and theological contexts—became a vivid experience for the many visitors who became fascinated with the various individualities of the singers. One woman decided she should stand with her back to a speaker in order to experience it as she would have when singing in a chorus, in which she heard most clearly the singer directly behind her.[9]

Thus, those experiencing the installation variously explored the shifting border between the individuality of the singers (including their random comments as well as the way they sang and breathed) and their coherence into what one might ordinarily have thought was "the" motet as a single unified sound-object. Here, technology allowed an intimacy of approach to each individual voice that would have been impossible or embarrassing in live performance. Why, then, did some of the visitors weep rather than express their reactions more neutrally? To be sure, their tears reflected the strong experience of this extraordinarily beautiful music, which many of them had not known or heard before, much less in such a memorable or intimate way. But perhaps they were also moved because they were exploring a deep penumbra of human experience that is rarely noted or accessible to observation: the shadowy and mysterious processes by which many separate voices—whether outside or inside the mind—meld into the shimmering appearance of oneness.

8 Controlling Dissonance

Whether four voices or forty, the character of a polyphonic composition depends crucially on how it uses dissonance between the voices, especially how those dissonances are resolved into consonances. The spectrum of such ways ranges from the greatest possible control over dissonance, using its careful preparation and resolution, as in Giovanni Palestrina's music, to the expressive force of unprepared dissonance, as in the works of Carlo Gesualdo. After 1600, these expressive devices became crucial to Claudio Monteverdi and others who created the new dramatic genre we now call opera. In astronomy, Johannes Kepler considered the polyphonic dissonances between the planets an essential aspect of the "harmony of the world" revealed by the new natural philosophy. Indeed, his search for consonances within the cosmic polyphony led him to discover a new relation between planetary periods and radii, the celebrated third law that Isaac Newton then used as a crucial step in establishing universal gravitation and the whole edifice of modern physics.

Whether in live performance, recording, or an installation, listeners experience the various ways and degrees to which polyphonic music coheres into a single whole. In the process, they experience different degrees of dissonance. Indeed, even among consonances there are subtle varieties of relative tension. Compared to the primal consonance of the octave (2:1), the next most consonant interval, the perfect fifth (3:2), already elicits a perceptibly greater felt tension. Trying to express this difference in his *Dialogues on Two New Sciences* (1638), Galileo Galilei memorably described the conflicting sensation caused by strings sounding a 3:2 ratio as producing "a tickling and teasing of the cartilage of the eardrum so that the sweetness is tempered by a sprinkling of sharpness, giving the impression of being simultaneously sweetly kissed, and bitten."[1] If this indeed describes a perfect consonance (a perfect fifth), one would need even more mixed images of pain and pleasure to describe outright dissonance. Box 8.1 outlines the changing hierarchy of consonances and dissonances from ancient Greek theory to about 1600.

Box 8.1
The hierarchy of consonances and dissonances

Starting already in ancient Greece with Pythagorean music theory, the primal consonance was the *octave* (such as C–c), sounded by two strings under the same tension having length ratios of 2:1. Next in order of consonance (and complexity of ratio) came the *perfect fifth* (C–G, 3:2) and then the *perfect fourth* (C–F, 4:3). Despite its simple ratio, though, by 1600 the perfect fourth was audibly functioning as an unstable harmonic interval; compared to the perfect fifth, perfect fourths seem to require resolution. By that time, what came to be called *imperfect consonances* had become increasingly important, in many ways more significant harmonically than octaves or fifths: the *major third* (C–E, 5:4), *minor third* (C–E♭, 6:5), *major sixth* (C–A, 5:3), and *minor sixth* (C–A♭, 8:5), called "just intonation." All other intervals were considered *dissonant*, to various degrees, such as the whole tone (C–D, major second, 9:8), the semitone (C–D♭, minor second, 16:15), the major seventh (C–B, 15:8), the minor seventh (C–B♭, 16:9), and the tritone (augmented fourth, C–F♯, 7:5). In particular, the ratios assigned the imperfect consonances and dissonances changed after 1600 with the gradual introduction of various kinds of *well-tempered* or *equal-tempered* scales.

Figure 8.1
The first four overtones above the fundamental note C, as observed by Marin Mersenne, showing their corresponding ratios (♪ sound example 8.1).

At the beginning of the seventeenth century, René Descartes and Marin Mersenne also described the basic phenomenon of overtones: a vibrating string produces not only its fundamental pitch but also a whole series of other pitches related to it (at the octave, the twelfth, and so forth, shown in figure 8.1; ♪ sound example 8.1). Thus, even a single string was in fact producing many pitches, not just one. This multiplicity bothered natural philosophers, narrowing the problem of polyphony down to a single tone: how could *one* string vibrate in many ways at once? Mersenne thought this represented a principle of harmonic pleasure: "The sound of any string is the more harmonious and agreeable, the greater the number of different sounds it makes heard at a time."[2] His answer seemed to take for granted that polyphony in general augments pleasure, compared to a single voice. Only much later did the full implications of the multiplicity of overtones in a single sound come forward as characterizing its timbre, as we shall see.

The correspondence between Mersenne and Descartes often included musical questions, as I have discussed elsewhere.[3] Descartes's first letter to Mersenne began by asking how one consonance could pass into another and yet "offer all the diversity of music," somehow uniting the multiple voices and intervals into a single harmony.[4] Considering the operations of the mind, Descartes was likewise troubled by how the many senses could ever coalesce into a single perceived whole, arguing that "the part of the body in which the soul directly exercises its functions is not the heart at all," as Aristotle had said, "or the whole of the brain." Drawing on anatomical studies that had reached new levels in the preceding century, Descartes noted that, for the most part,

all the other parts of our brain are double, as also are all the organs of our external senses—eyes, hands, ears and so on. But in so far as we have only one simple thought about a given object at any one time, there must necessarily be some place where the two images coming through the two eyes, or the two impressions coming from a single object through the double organs of any other sense, can come together in a single image or impression before reaching the soul, so that they do not present to it two objects instead of one.[5]

Because the pineal gland, in "the innermost part of the brain," is *not* double, Descartes argued that it ought to be where the doubled sounds or images could come together, the *sensorium commune* or "common sense" postulated from Aristotle on.[6] Further, Descartes (mistakenly) thought that this gland was not present in animals, hence was appropriate for the seat of the human soul.[7] Though his suggestion soon came under criticism, he was among the first to bring to light the problematic relation between the doubled brain hemispheres (and the other doubled sense organs) and the seeming unity of perception. Similarly, the problem that polyphony can be apprehended in some kind of unity parallels Descartes's concern for the unity of soul in the face of multiple sensory modalities.

As noted in the previous chapter, increasing the number of voices results in an increasing pressure on composers to use consonance to avoid cacophony: with ever more voices, each of them must be harmonized with so many others that it is safer to avoid conflicts by using more consonant than dissonant intervals. The fundamental issue emerges more clearly when one returns to a smaller number of voices.

For example, consider the work that became the *beau ideal* of Roman Catholic liturgical music, Palestrina's *Pope Marcellus Mass* (*Missa Papae Marcelli*, 1562), which cardinals acclaimed already in 1565 with "the greatest and most incessant praise."[8] Palestrina's mass for six voices projects an aura of immaculate serenity through its pristine counterpoint, qualities that led to the long-standing (though finally unjustified) myth mentioned earlier that this mass singlehandedly "saved" polyphonic music from condemnation at the Council of Trent. Looking more closely at Palestrina's writing in his Kyrie I, for example, one realizes that its otherworldly calm comes from its exacting control of dissonance. To create a predominantly consonant texture, Palestrina strictly limited the motion of any one voice, never allowing leaps of a major sixth, for instance, and preferring scalewise motion. In this way, he regulated the degree of dissonance of each voice considered as a melody in itself, unfolding over time, by making sure that each dissonance was carefully *prepared* and *resolved*.

Thus, Palestrina never allowed the voices to arrive at or leave a dissonance by skip, but only by step (see figure 8.2; ♪ sound example 8.2).[9] By so doing, he made the appearance (and disappearance) of the dissonance as gradual as possible, eliminating any sudden dissonance: he never allowed an unprepared dissonance, for instance. The voices all imitate a subject beginning with a rising fourth (traditionally considered a perfect consonance, though one that by then functioned almost as a dissonance) that evokes a subdued exaltation, followed by soft downward motion. In this way, dissonance is "tamed" to give variation and tension so as to alleviate unremitting consonance without ruffling the music's exquisitely calm surface. Palestrina's music is *diatonic*, meaning that it uses the unaltered sequence of notes in its mode (in modern terms, the pure C major given by the white keys of a piano); he avoids any sharps or flats ("accidentals"). As Philip LeCuyer has pointed out, Palestrina also arranges the overlay of the words so that at one point (the end of measure 21) all the syllables of the text ("Kyrie eleison") are pronounced simultaneously, adding a further level of linguistic polyphony.[10]

In contrast to Palestrina's careful preparation, an unprepared dissonance gives a sudden stab of expression. Consider, for example, the end of William Byrd's Agnus Dei from his *Mass for Four Voices* (written about 1592), which sets the final words of the liturgy, "*dona nobis pacem*" ("grant us peace"), to a haunting series of suspensions, meaning that an initially consonant note becomes dissonant because of the subsequent motion of another voice. When suspensions are reached by a leap in one of the voices, each successive dissonance conveys a wave of pain that is revealed through the successive motion of the voices (figure 8.3; ♪ sound example 8.3). For instance, the soprano D in measure 43 is initially consonant with the notes around it (G and B♭) but suddenly becomes dissonant in relation to the bass leaping to E♭ at the beginning of the next measure (44). Dissonance is, after all, relative and subject to the changing context of the notes around it. By contrast, Palestrina never allows one of his voices to leap in a context where a dissonance would thus suddenly appear.

Writing during a time of constant war, as a Roman Catholic practicing a prohibited faith in an adamantly Protestant land, Byrd well understood the elusiveness of peace. His plea rises from successive waves of dissonance and resolution, each preparing another wave to come (measures 43–54), all reiterating the primordial discord of a semitone, here D against E♭, A♭ against G, B♭ against A. Yet when it seems that these ebbing waves of dissonance may never end, Byrd's final four measures draw from its interweaving contrapuntal lines a perfect and utterly serene cadence on G. Where Machaut, Ockeghem, and Josquin ended their works with bare fifths (such as the D–A of Machaut's Kyrie I), Byrd ends on the sweetness of a full major triad, G–B–D; Palestrina also closes his Kyrie I with a major triad (C–E–G). After the dissonant suspensions of Byrd's preceding measures (and the resultant minor triads), the final major triad not only symbolizes but actually conveys "the peace that passeth all understanding."

Figure 8.2
Giovanni da Palestrina, Kyrie I from *Missa Papae Marcelli* (1562) (♪ sound example 8.2). Note that all dissonances between voices are carefully prepared and resolved, always approached and resolved by stepwise motion. These dissonances are often passing notes (between two consonances) or neighboring notes to a consonance. Palestrina restricts himself to the use of only those notes lying in what we would now call C major (with the exception of one possible accidental, the F# in measure 4). Text: "Lord, have mercy."

Figure 8.3
William Byrd, the end of the Agnus Dei from *Mass for Four Voices* (♪ sound example 8.3). Text: "Grant us peace." Note the dissonant suspensions on the syllable "*do-*" of *dona*.

During the fifteenth and sixteenth centuries, many polyphonic works used various amounts of dissonance for expressive purposes. Among myriad examples, the compositions of Carlo Gesualdo, prince of Venosa, are especially vivid. The lurid details of his life—killing his wife and her lover, flagellating himself in repentance—should not obscure the way he used dissonance with unparalleled freedom to evoke the deepest expression of his texts. Among his sacred works, his 1611 setting of the *Tenebrae* (the solemn service for Maundy Thursday commemorating Jesus's abandonment by his disciples) is very far from the calm sound-world of Palestrina. In Gesualdo's motet *Tristis est anima mea*, the biblical text evokes Jesus's dark vigil in the Garden of Gethsemane: "My soul is sorrowful even unto death." When Gesualdo sets the text "you shall run away [*vos fugam capietis*]," the initial "*vos*" defiantly identifies *you*, the disciples who flee, using rapid, vigorous figures; as Jesus continues "I will go to be sacrificed for you," the texture is hesitant, interrupted by silences. Gesualdo uses one unprepared dissonance after another, forming a halting, broken series of harmonies, so beautiful and painful, suggesting the mortal anguish with which Jesus was confronted at the moment in which he realized the full magnitude of his sacrifice (figure 8.4; ♪ sound example 8.4). These harmonies are *chromatic*, meaning that they involve melodic motions by a semitone (such as from C# to C natural), the dissonant interval that (since ancient times) was associated with color (*chroma*) and contrasted with the unaltered "purity" of diatonic motion (such as in Palestrina).

Here, the use of chromaticism raised dissonant possibilities to their heights. Palestrina had not allowed any voice to move by a semitone, apart from those required within its modal scale, nor permitted the voices to form semitone dissonances with each other. Byrd used only the latter, Gesualdo both, of these devices. In different ways, both composers were part of a larger spiritual movement toward greater expressive intensity. Ignatius Loyola's *Spiritual Exercises* advised the devout to advance their spiritual lives by picturing the most intense (and harrowing) moments of sacred history in vivid detail; Byrd and Gesualdo shaped dissonances that virtually forced their listeners to feel the anguished prayer for peace, the desolation of the abandoned Redeemer. In contrast to Byrd's quiet subtlety, Gesualdo dramatized to the highest degree. Each of them aspired to different visions of music—contemplative tranquility versus dramatic intensity.

These expressive dissonances, more and less prepared, were crucial techniques in the emergence of a new kind of music conceived as essentially dramatic. This new music sought the power to move the passions rather than a dispassionate evocation of beauty. During the sixteenth century, many explorations of expressivity in madrigals and songs led up to the earliest operas (as we now call them), works that consciously sought to revive and even excel the fabled powers of ancient Greek drama.[11] In the last half of that century, aristocratic patronage of these new projects grew in such circles as the Florentine Camerata around Count Giovanni de' Bardi, the patron of Galileo's father Vincenzo Galilei, a lutenist and composer who became fascinated with ancient Greek music.[12] Vincenzo's *Dialogue on Ancient and Modern Music* (1581) was "surely the most influential music

Figure 8.4
Carlo Gesualdo, *Tristis est anima mea* from *Tenebrae* (1611) (♪ sound example 8.4). Text: "You shall run away, and I will go to be sacrificed for you."

Figure 8.4 (continued)

treatise of the late sixteenth century."[13] In its polemic for a new kind of expressive, dramatic music, Vincenzo pointedly attacked contemporary polyphony. Though madrigals used various expressive devices to dramatize their texts (comparable to some Gesualdo used in the example above), Vincenzo found them laughable. In his dialogue, his protagonist Bardi (named after his patron) made fun of the madrigalists' practices of setting texts about "fleeing or flying" merely "by making the music move with such speed and so little grace that just imagining it is enough." Likewise, to set words like "swoon" or "die," the madrigalists "make the parts suddenly fall silent so abruptly that instead of inducing in listeners corresponding affections, they provoke laughter and contempt."[14]

In Vincenzo's judgment, though, the fundamental problem was not so much the inadequacy of these attempts at textual expression as the nature of polyphony itself, as practiced in his time. His objection was not to the plurality of voices as such, but (ironically) to what he considered their unremitting and inexpressive consonance. Here he reflected on the need we discussed above for polyphonic voices to seek consonance in order to control their dissonances, hence the need for rules such as Palestrina observed and Vincenzo's teacher Gioseffo Zarlino codified.[15] Vincenzo conceded that "there is no one who would not consider these rules optimal and necessary for the simple delight that the hearing takes in chords and their variety. But for the expression of ideas and affections they are pernicious, because they are good only to make the harmony varied and full, which does not always—rather never—suit the expression of any conceit of a poet or orator."[16] In his view, polyphony generally does not do justice to that kind of drama. Instead, he quotes the critique "the divine Plato" made of polyphony (which we discussed in chapter 1) in order to admonish contemporary composers that one should sing "in unison and not with consonance."[17]

To be sure, one can find many examples in which composers used beautifully the very devices for word painting that Vincenzo derided. His sarcasm may perhaps be understood as a complaint about the widespread use of those formulas in mediocre or uninspired ways. Still, he was serious about seeking a more dramatic form of declamation than what he thought possible in polyphonic music, with its overarching need to coordinate its various voices toward consonance, thereby diluting its dramatic power. In his view, the goal of "modern" contrapuntal music "is only to delight the hearing, while that of the ancient was to lead others by its means into the same affection as one felt oneself."[18]

To remedy this basic problem, Vincenzo wanted to revive the practice in which "the ancients sang their tragedies and comedies to the sound of the aulos and the kithara" (see figures 1.6 and 8.5). Thus, "when an actor sang in unison—and not in consonance, as we said and proved—with the instrument—whether this was the aulos, the kithara, or another instrument—this was the way he was best heard by those around him and his voice tired the least."[19] The accompanying instrument helped support the singer's pitch and rhythm, so that he could articulate "what suited the character of the ideas that the actor sought to express with the words."[20]

Figure 8.5
A man and woman in a chariot, accompanied by another woman and a kithara player (ca. 540 BCE). Apollo was often depicted playing the lyre..

That was the crux: rather than arbitrary or forced associations of the text with musical devices, the ancients clothed their texts with music "that suited the action of the personage." Vincenzo argued that, compared to a polyphonic ensemble, a single voice could better show the distinctions between "a prince talking with his subjects or vassals, or a suppliant pleading, how a furious or excited person speaks, how a married woman, a girl, a mere tot, a clever harlot, someone in love speaking to his beloved when he is trying to bend her to his will, how someone who laments, or one who cries out, how a timid person or one exulting in joy sounds."[21] This *monody* (as it later came to be called) thus was not purely monophonic (like chant) but rather a single vocal line sung to instrumental accompaniment that was understood to have a subsidiary, supporting role, rather than being a rival voice. Still, this accompaniment was generally improvised on a bass line provided by the composer (the continuo we encountered in Striggio's forty-voice motet), offering rich possibilities of texture and ornamentation that would reflect the character and situation at hand.

When the dreams of Vincenzo and his friends came to fruition in the earliest operas, monody took a prominent place indeed. Growing out the princely entertainments that Leonardo da Vinci crafted, court masques and *intermedi* (interludes) took on the trappings of ancient drama, often drawing on Greek mythology. Here political considerations once again drew on the prestige of polyphony. Having used Striggio's forty-voice compositions to help gain the status of grand dukes, twenty years later the Medici also employed polyphony to add special éclat to an important dynastic event, the wedding of Ferdinando de' Medici and Christine of Lorraine in 1589.[22]

Bardi took a leading role in conceiving spectacular *intermedi* for the occasion, which he took as an opportunity to stage ancient scenes with modern grandeur. For instance, the first *intermedio* was an elaborate presentation of "The Harmony of the Spheres" closely based on Plato's depiction of the cosmic spindle of Necessity we considered in chapter 1. In a sumptuous stage design by the mannerist artist Bernardo Buontalenti, the three Fates and eight planetary Sirens sang to honor the nuptial pair, using more singers than in the ancient text in order to perform many-voiced compositions by Cristofano Malvezzi and others (figure 8.6; ♪ sound example 8.5). Indeed, the wedding intermedi featured choruses of six,

Figure 8.6
Agostino Carracci, *The Harmony of the Cosmos* (1589–92), showing the spindle of Necessity (top center), the three Fates at her feet, surrounded by many more singers than the eight planetary Sirens in Plato, presumably to reflect its commemoration of the stage design for the first *intermedio* in the sumptuous Florentine wedding celebrations for Ferdinando de' Medici and Christine of Lorraine (1589). This design was based on the work of Bernardo Buontalenti (see figure 1.7).

twelve, eighteen, even thirty parts, though these aimed more for homophonic grandeur than truly independent counterpoint.[23]

Beside such choral showpieces, these *intermedi* included dramatic monody, beginning with Armonia (Harmony) descending from the highest sphere to greet the "the new Minerva and mighty Hercules" at their nuptials. Going even further, one of the greatest of the early operas, Claudio Monteverdi's "fable in music" (*favola in musica*) *L'Orfeo* (1607) explicitly put La Musica—Music herself—onstage in the prologue to declaim her purpose: "I am Music, who can move even the most icy mind." Expressions of feeling and character permeate the story, which turns dramatically on whether Orpheus, the archetypal expressive musician, can master his own passions as well as he can move those of others.

But Monteverdi himself was far from doctrinaire about using polyphony, to whatever extent he may have known or approved of Vincenzo's writings. To be sure, Monteverdi used monody wonderfully, as when Orpheus movingly bids farewell to Earth and heaven before descending into the underworld to rescue Eurydice (figure 8.7; ♪ sound example 8.6). Nevertheless, Monteverdi also employed polyphony on many occasions in this work, both creating grand statements and expressing a variety of emotions and situations. For instance, he crafted an imposing five-part chorus to praise Orpheus's victory over the infernal powers, a chorus whose text closely imitates Sophocles's great choral ode from *Antigone*: "Many are the wonders of the world, but none more wonderful than man" (figure 8.8; ♪ sound example 8.7).

Throughout this opera, Monteverdi freely used both monody and polyphony as suited his dramatic purposes; for the most sensitive and expressive moments he consistently chose monody, using polyphony for expressions of a whole group, such as the wedding chorus sung by the assembled nymphs and shepherds, their communal outpourings of joy or grief. Other notable polyphonic moments in the opera include many purely instrumental passages (often dance-based) and Orfeo's final duet with his father, Apollo, as they ascend to the skies, leaving the sad Earth behind.

Figure 8.7
Orfeo's monody from act 2 of *L'Orfeo* (♪ sound example 8.6) by Claudio Monteverdi. Note that the bottom line indicates the continuo, a bass line on which various instruments (such as lutes, theorbos, harpsichords, or organ) would then improvise triadic accompaniments. Text: "Farewell, Earth, farewell, Heaven, and Sun, farewell."

Figure 8.8
The opening of the final chorus from act 3 of *L'Orfeo* (♪ sound example 8.7). The bottom line indicates the continuo (bass). Text: "Nothing undertaken by man is attempted in vain, nor can Nature defend herself against him."

Public letters from 1605–1607 by Claudio and his brother Giulio Cesare Monteverdi explained his new music and replied to criticisms that it violated accepted rules of counterpoint. To do so, they defined a "first practice" (*prima prattica*), exemplified by such contrapuntal composers as Josquin and codified by Zarlino, a practice that "turns on the perfection of the harmony, that is, the one that considers the harmony not commanded but commanding, and not the servant, but the mistress of the words."[24] In contrast, they called the "second practice" (*seconda prattica*) a dramatic art pioneered by such masters as Cipriano de Rore and Adrian Willaert, an art that "turns on the perfection of the 'melody,' that is, the one that considers harmony commanded, not commanding and makes the words the mistress of the harmony."[25] By making this distinction, the Monteverdis indicated that they did not repudiate the first practice (or wish to engage in fruitless polemics with the advocates of that style) but rather wanted to add the possibilities of "a second practical usage," freer in its use of dissonance that would allow more expressive settings of texts.

Soon after these new exploitations of the relative powers of polyphony and dramatic monody, Johannes Kepler used polyphony as an important conceptual element in his new astronomy. Thus, though there seems a great distance between these first operas and Kepler's astronomy, both were concerned with the status and significance of polyphony as well as of dissonance. Indeed, Kepler's *Harmonices mundi libri V* (*Harmony of the World*, 1619) is one of the most extraordinary defenses of polyphony ever written. Kepler had thought a great deal about music as well as astronomy. Traveling in 1617 to save his aged mother from prosecution as a witch, Kepler read Vincenzo Galilei's *Dialogo* "with the greatest pleasure [*summa cum voluptate*]," though he vehemently disagreed with its ideas about tuning, showing his sensitivity to these issues.[26] As Kepler reconsidered and revised the ancient topic of cosmic harmony, he asserted that the universe was essentially polyphonic, not monophonic. He based his interpretation on the common belief of his time that ancient music, being monophonic, likewise had a monophonic conception of the "music of the spheres." As I have discussed elsewhere, Kepler expressed his strong feeling for contemporary polyphonic music by referencing works by Orlando di Lasso, particularly his motet *In me transierunt* (1562).[27] Kepler drew attention to Lasso's expressive dissonances, especially the way he uses the semitone to "sound plaintive, broken, and in a sense lamentable," suiting this motet's penitential text from the Psalms: "Thy wrath has swept over me; thy terrors destroy me."[28]

Kepler used the most accurate contemporary astronomical observations by Tycho Brahe to compile what he thought were the real songs of the planets (figure 8.9; ♪ sound example 8.8). He thought carefully about the resultant polyphony that would come from the simultaneous performance of those songs, each reflecting the period of its planet and the eccentricity of its orbit around the Sun (meaning the ratio by which it departs from perfect circularity). In contrast to the Ptolemaic system of the ancients, Kepler's Earth is no longer stationary and mute, but has its own song as it circles the Sun (shown in figure

Figure 8.9
Johannes Kepler's planetary songs (including also the Moon); note the song of the Earth (Terra), spelled *mi fa mi* in the notation of the time (♪ sound example 8.8).

8.9), which is just a wailing semitone: "the Earth sings *MI FA MI*, so that even from the syllable you may guess that in this home of ours MIsery and FAmine [*MIseria et FAmes*] hold sway."[29]

Kepler was keenly aware that these planetary songs, when combined simultaneously, resulted in a notably dissonant cosmic polyphony, as you can hear for yourself in ♪ sound example 8.8. Perhaps ironically, he emphasized the "marital" conflicts between the Earth and Venus, whom he conceived as "masculine" and "feminine" planets whose songs notably conflict (the Earth's semitone versus Venus singing within a diesis, a quartertone). In a further irony, Kepler praised the specifically sexual feeling of polyphonic cadences, thus inverting the old objection that polyphony tends to sexual titillation.[30] Indeed, his sexual interpretation of these cadences depended absolutely on their polyphony: his description of the interplay of two voices reads their various cadences in terms of masculine and feminine orgasms, which he also applied to the relative positions of two planets. Taking a still larger view, Kepler presented these cosmic dissonances as expressing and even explaining the presence of evil in the world, the "misery and famine" he saw all about him, the constant wars of his times. Still, he investigated what planetary configurations might lead to a more consonant cosmic sound, at least at some auspicious moments.

Though his search was inconclusive, in its course he discovered a simple (but previously unknown) proportionality between the square of a planet's orbital period and the cube of its mean distance from the Sun, the extraordinary relation later called Kepler's third law. In his *Harmony of the World*, he relates that he had been seeking the exact relation between these quantities for twenty-two years, since his first book, *Mysterium cosmographicum* (*The Cosmic Mystery*, 1591). But only through using "the observations of Brahe, by continuous toil for a very long time, at last, at last, the genuine proportion of the periodic times to the proportions of the spheres" became clear to him.[31] So important was the moment of discovery that he recorded it as precisely May 15, 1618, when (after a false start he dated to March 8) this idea, "adopting a new line of attack, stormed the darkness of my mind." He reached this momentous result both through his "contemplation of the celestial harmonies," their detailed polyphonic interrelations, and Brahe's precise observations; the "sesquialterate proportion" he found between periodic times and planetary mean distances is the same mathematical and musical term that describes a perfect fifth (the "sesquialterate ratio," 3:2).

Kepler's exploration of cosmic polyphony did not stop at this point. He remained troubled that the cosmic polyphony was so dissonant and the planetary orbits so incommensurable that no "final cadence," no consonant cosmic harmony could be expected that would appropriately end the universe, as Scripture (and orthodox Christianity, whether Catholic or Protestant) seemed to demand. Kepler's never-repeating, dissonant cosmic polyphony may have been the first and most enduring expression of the universe revealed by the new natural philosophy he pioneered. Yet from that dissonant polyphony emerged subtle but all-important harmonic laws. When Isaac Newton "used to say he believed Pythagoras's Musick of the Spheres was gravity," he indicated how important he considered the celestial polyphony that led to Kepler's discoveries and then to his own establishment of the law of universal gravitation, on which all subsequent physics depends.[32]

9 Contrapuntal Science and Art

The new natural philosophy of this time, including Kepler's polyphonic cosmology, was preceded and paralleled by the development of a science of counterpoint that codified the rules of polyphonic practice. During the sixteenth and seventeenth centuries, the writings of Gioseffo Zarlino, Thomas Morley, and Johann Joseph Fux presented this ongoing and changing endeavor in different ways, addressing not only professional composers and performers but also a broad audience of serious amateurs. This increasingly codified contrapuntal science aspired to render polyphony into something more intelligible and learnable than merely the precepts of a craft. My concern here is not so much the details of those principles as the nature and significance of this new musical science.

To be sure, music had been a science since Greek times as part of the quadrivium of liberal arts. In that context, *musica* (or "harmonics," as it also came to be called) concerned the fundamental mathematical ratios that were the elements of which all music was made, whether the cosmic *musica mundana* or the audible *musica instrumentalis*. Augustine had famously defined music as "the science of making melody well [*scientia bene modulandi*]."[1] To this, the *Enchiriadis* treatises and Guido d'Arezzo (among others) had already included material about the making of polyphony, based on those fundamental mathematical elements.[2] Only in the thirteenth century, though, did the terms *theoria* and *practica* start to be used to describe ancient harmonics and the newer practices of counterpoint, respectively.[3]

The emergence of didactic works about practical music represented a change from older traditions. In the Middle Ages, apart from singers in ecclesiastical choirs, most musicians were professionals, from minstrels and wandering players eking out a living in taverns to those employed by princely courts. Many of those paid musicians thought of themselves as part of a guild that guarded the secrets of their craft so as to protect the livelihoods of the members. Thus, musicians were not eager to publish their secrets for others to learn.[4] During the fifteenth century, education in the liberal arts (including music) was increasingly extended to more people, often of higher social standing.[5] Even more important were social factors: entertainment at noble houses usually involved music-making by the guests (rather than paid players), so that a passable knowledge of musical elements and practice (as well as of dance) became essential for anyone who wished to take part in social life, especially if they wished to shine. Serious amateur musicians were eager to buy works that would help instruct them in these all-important social skills.

Johannes Tinctoris's *Liber de arte contrapuncti* (*The Art of Counterpoint*, 1477) was among the first treatments to call counterpoint an "art," a term implying that it was comparable to the ancient liberal arts, particularly the mathematical elements of music. He specifies that this contrapuntal art should be described by "reason and science" as well as by "the highest constant effort of practice," for which he provides examples from the great composers of his time (as noted in chapter 6).[6] Elsewhere, he connected the "new art" of his time and the science of arithmetic to "the greatest of musicians, Jesus Christ," who "in duple proportion made two natures one," a rare echo of Augustine's analogy between the octave and the Incarnation we discussed earlier.[7]

The humanist revival of ancient Greek thought in the fifteenth century sparked what one scholar called a veritable "mania for music theory."[8] Among Tinctoris's followers, Franchinus Gaffurius resurrected the Greek word "theory" in what was the first printed book on "music theory," *Theoricum opus musice discipline* (1480), a digest of Boethius's musical writings (until that point still only available in manuscript), and then in Gaffurius's *Theorica musicae* (1492). Until that time, the terms *theoria* or *theorica* had mainly been restricted to works describing astronomy (the *theorica planetarum*, "theory of the planets") and music. Thus, the concept of "theory," so important for the subsequent development of natural science, seems to have come to it from music and astronomy in the quadrivial tradition.[9] Gaffurius concluded that "at first, music consisted by its nature in practice, but now it is examined in art. Now finally it is a science [*ratio*]."[10]

Gaffurius devoted his *Theorica musicae* to the mathematics underlying music; he wrote a separate *Practica musicae* (1496) to describe the system of notes, rests, and modes used in counterpoint; no other work "achieved as great an influence upon the musical thought of sixteenth-century Western Europe."[11] He devoted a great deal of attention to the rhythmic structure of polyphonic music, which he limited to rational proportions, though recognizing the importance of irrational proportions in geometry.[12] For the numerical structure of the pitches, he presented eight rules that govern the composition of counterpoint, such as using perfect consonances to begin and end a work and interspersing imperfect consonances and dissonances in order to avoid the parallel motion of the voices, especially what he called the "mandatory" avoidance of parallel fifths or octaves.[13] He did not explain this prohibition (to which we will return), first stated by Johannes de Grocheio about 1300. More practically, Gaffurius warned singers to avoid "absurd loud bellowing," "unsightly opening of their mouths," "extravagant and indecorous movement of the head or hands," as well as "excessive vibrato."[14] He also advised composers "to adapt the melody in its sweetness to the words of the song, so that when the words concern love or a longing for death or some lamentation, he will articulate and arrange doleful sounds so far as he can, as the Venetians are wont to do," thus reflecting contemporary practices of expressive declamation, whose growth we have been tracking.[15] Gaffurius ended this book by inviting "the most skilled mathematicians and musicians" to judge his work, for he considered his practical text to involve both fields.[16]

Gioseffo Zarlino, in his *Le Istitutioni harmonice* (The harmonic institutions, 1558) provided the first unified treatment of *musica theorica* and *musica practica* in one work. In so doing, he created a new synthesis in which counterpoint took a central role, now to be understood not merely as a "practice" but as a science equal in rigor and scope to the older teachings of harmonics, a "science that deals with sounds and tones."[17] Zarlino had studied with the Venetian master Adrian Willaert, whom he praised as "truly one of the most rare intellects that has ever practiced the art of music, ... a new Pythagoras," and whom he eventually succeeded as maestro di capella at St. Mark's.[18] Zarlino took upon himself his teacher's quest to rescue music from the "extreme baseness" he considered it had reached, "having retained neither part nor vestige of any of that venerable weight which it used to possess in ancient times."[19] He saw himself "gathering various things from the old masters," such as Josquin des Prez (whom he instances along with Willaert), "and also discovering something new" as well as "disclosing [music's] secrets."[20]

Zarlino began by connecting music with human intelligence, "something that barely distinguished him from the angels," so that God "created him with his face turned up to the sky, where the seat of the Lord is."[21] Hearing extends farther than any other sense "and is much more useful in the acquisition of science and in intellectual judgment. Therefore, it follows that hearing is truly more necessary and better than the other senses," giving rise to "the science of music."[22] Zarlino also considered "our science of music" to parallel Aristotle's "true science of natural philosophy"; as "part of the mathematical sciences ... [music] exceeds in certainty the other sciences, and holds the first rank of truth."[23] Some Pythagoreans even held "that music holds primacy among the liberal arts and some of them called it *encyclopedia* meaning ... 'circle of sciences'; so that music, as Plato says, embraces all disciplines."[24]

Zarlino argued that the student "must accompany [music] with the study of theory [*speculatiua*]" as well as "the other sciences."[25] He reorganized the traditional three categories of music into two: "animistic music" (comprising Boethius's *musica mundana* and *musica humana*) and "organic" (Boethius's *musica instrumentalis*), which he divided further into "natural" (vocal) and "artificial" (instrumental) music. Zarlino's "perfect musician" would combine skill in musical practice (composition) with understanding "the logic belonging to [musical] science."[26] In so doing, Zarlino installed "the art of counterpoint" alongside the mathematical theory of ratios, here using the term "art" interchangeably with "science."

The central Part III of his work presents his treatment of practical counterpoint in terms of reasoned rules. Compared to Gaffurius, Zarlino gave a far more extensive account of these rules and their justification. For instance, parallel fifths and octaves should be prohibited because there is no "diversity of melodic motion, for both parts proceed by the same intervals. … Just as a painting in many colors pleases the eye more than a monochrome, varied consonances and melodic movements please the ear more than the simple and invariant, and therefore a diligent composer uses variety in his work."[27] Zarlino thus used pleasure as a guiding criteron, rather than relying on purely mathematical arguments. He also offered practical guidance to the composer on choosing an appropriate subject (melodic motive or cantus firmus); he was concerned with singability and elegance.

Like Gaffurius, Zarlino believed that "a musical composition shall complement the text, that is the words. With gay texts it should not be plaintive, and vice versa."[28] Zarlino was conservative in his tastes; he specifically critiqued "chromatic compositions by certain moderns," presumably composers like Nicola Vicentino, who advocated dividing semitones into quartertones in order to revive the ancient enharmonic genus of music:[29]

Here is how they justify this. The voice is capable of forming any interval, and it is necessary to imitate ordinary speech in representing the words as orators do and ought. Therefore it is not inappropriate to use all these intervals to express the ideas contained in the words, with the same accents and other effects we use in conversation, so that the music might move the affections. … They say we must imitate orators if our music is to move the affections. Yet I never heard an orator use the strange, crude intervals used by these chromaticists.[30]

This was an argument later taken up by Giovanni Maria Artusi to criticize the expressive dissonances of Gesualdo and Monteverdi. Distressed by these "strange, crude intervals" and their jarring theatricality, Zarlino ended by hoping that "some noble spirit" would follow him by "laboring in our science and raising it to the ultimate point that I have suggested."[31]

Zarlino's treatise indeed had immense influence in the succeeding decades, though his rebellious student Vincenzo Galilei attacked his treatment of ancient music and raised the banner for the kind of conversational declamation that Zarlino thought so ugly. Nonetheless, Thomas Morley's *A Plaine and Easie Introduction to Practicall Musicke* (1597), the first printed book in English about music, is content to rely on Zarlino as his major authority. A student of William Byrd, Morley was an outstanding composer. Indeed, his book reveals a whole social subtext to the study of counterpoint. Cast as a dialogue between Polymathes ("learned in many fields"), his brother Philomathes ("lover of learning"), and the Master (Morley himself), the seemingly arcane subject of counterpoint turns out to be a locus of social anxiety. To his brother, Philomathes confesses the ultimate social nightmare: "Supper being ended and music books (according to the custom) being brought to the table, the mistress of the house presented me with a part earnestly requesting me to sing; but when, after many excuses, I protested unfeignedly that I could not, every one began to wonder; yea, some whispered to others demanding how I was brought up … [so that] the whole company condemned me of discourtesy."[32] The very next day, Philomathes hastens to seek out his old music master in order "to make myself his scholar" again. After condescendingly wishing him good luck (and rubbing in his superior virtue), his brother goes off to hear some mathematical lectures. The master is surprised to see his old student return, "for I have heard you so much speak against that art [music] as to term it a corrupter of good manners and an allurement to vices," bringing up the stinging accusation of music-hating we have considered in the context of the polyphonic controversies. Philomathes has to grin and bear his master's rebuke; he confesses that he "learned nothing in music before" and begs the master to "begin at the very beginning and teach me as though I were a child."[33]

They begin with the elements of musical notation, explained in great detail, making the book a wonderful resource for those who wish to learn from the inside the musical world of Elizabethan England. The master gives many examples and repeatedly tests his repentant student on singing a part of a contrapuntal composition, the very skill he so woefully lacks. Thus, the needed musical literacy was above all the ability to understand the notation and performance of complex contrapuntal works, such as were evidently sung after many a supper. The larger context is indicated by the book's beautiful title page (figure 9.1), in which the four liberal arts and the great authors of geography, astronomy, and music together form a glorious garland that encircles and exalts "practicall musicke."

The lessons bear satisfying fruit; at the beginning of the book's second part, Philomathes tells the master that he recently out-sang his brother, who previously had mocked his lack of knowledge but now "I might set him at school." Hearing someone praised as "the best Descanter to be found," Philomathes is eager to learn descant so that he too can garner such admiration.[34] The master then puts him through a lengthy series of examples to teach him how to compose and sing (even extempore) another part to a cantus firmus. At first, they work through simplified cases, one long note against another. Philomathes learns by experience the forbidden sound of parallel octaves or fifths, including its slang name, "hitting the octave on the face."[35] Gradually, they reach the heights of canon and complex counterpoint. In the third and final part, brother Polymathes joins them, after returning from a lecture on Ptolemy's astronomy; his superior pretensions suffer when his counterpoint is revealed as decidedly inferior, the result of his having relied on a teacher named Bold, who was evidently more confident than knowledgeable. All the while, the brothers are working out quite complex examples in four, five, and even six voices under the master's guidance; one gathers that, in cultivated circles of the time, an acceptable knowledge of counterpoint had to be surprisingly advanced. To help his students, the master presents helpful tables of commonly encountered situations in counterpoint, though his instruction tends more to address particular problems rather than giving sweeping precepts. Reading Morley's book, one has a chance to look over his shoulder as he comments on various difficulties and offers alternatives.

Figure 9.1
The title page of Thomas Morley, *A Plaine and Easie Introduction to Practicall Musicke* (1597). The four liberal arts (counterclockwise geometry, arithmetic, music, and astronomy) are depicted at the bottom; above them are images of great authors in those subjects: Polybius, Strabo, Marinus, Ptolemy, Aratus, and Hipparchus.

In contrast, Johann Joseph Fux's *Gradus ad Parnassum* (*Steps to Parnassus*, 1725) transformed this long tradition of more informal contrapuntal instruction into a precise method, which "has continued to be the single most influential force in counterpoint teaching through to the beginning of the twenty-first century."[36] His title proposes a series of graduated steps by which the learner can climb Parnassus, the summit of musical art and mythical home of the muses (figure 9.2). Though it seems paradoxical that the peak of compositional inspiration could be approached methodically, Fux's work was important in the education of Haydn, Mozart, Beethoven, Schubert, and Brahms, among many others.

Though Morley's *Introduction* was addressed in the first instance to anxious amateurs, Fux was writing for a more determinedly vocational audience, specifically "young persons … who have fine talents and are most anxious to study; however, lacking means and a teacher, they cannot realize their ambition, but remain, as it were, forever desperately athirst."[37] This description certainly applied to Haydn and not a few of the other great composers who studied Fux closely. Fux's book presents a dialogue between Aloysius, representing "Palestrina, the celebrated light of music," and Joseph (Fux's own middle name). Thus, Fux stations himself as the student of Palestrina no less than the exponent of the *stile antico*, the "ancient style" of which he considered Palestrina to be the exemplar.

Indeed, the greatness of Fux's approach was to lay bare the most fundamental elements of counterpoint and present them tersely, then showing how they are to be applied. Before the dialogue itself commences, his Book I presents the *pars speculativa*, a summary of the Boethian account of musical ratios and intervals that starts by listing "celestial music [*Musicam caelestem*]" as one of the genera of music, along with "terrestrial, natural, artificial, historical, etc."[38] Though Fux qualifies this theoretical introduction as "brief" (forty-two pages out of almost three hundred), he includes quite a bit of mathematical detail from ancient musical theory. He then lists all the intervals, distinguishing perfect and imperfect consonances from dissonances, defines the three kinds of motion in music (direct, contrary, or oblique), and then presents the four fundamental rules on which all counterpoint depends (see box 9.1). In his notes on Fux, Beethoven remarked that these rules were, "strictly speaking, only two in number" and Padre Martini (Mozart's teacher) reduced them to only one. "the only progression forbidden is the direct motion into a perfect consonance."[39] By comparison, Zarlino offered twelve rules and Morley did not enumerate his. This represents not only a great pedagogical simplification but the formulation of laws that, in their generality, were comparable to the "summary laws of nature" that Francis Bacon proposed as the highest aspiration of the new philosophy, such as Isaac Newton's laws of motion, celebrated throughout Europe by the time of Fux's book.

Figure 9.2
The frontispiece of Johann Joseph Fux's *Gradus ad Parnassum* (1725), showing the author being crowned with laurel by Apollo, as the Muses look on; the neat steps contrast with the rocky slope. The eagle insignia indicates that the work was published under imperial patronage. Note that images of instruments and theatrical masks of tragedy and comedy surround the display, rather than the representations of the liberal arts and ancient sages on Morley's title page (figure 9.1).

Box 9.1
Fux's four rules of counterpoint

First rule: From one perfect consonance to another perfect consonance one must proceed in contrary or oblique motion.

Second rule: From a perfect consonance to an imperfect consonance one may proceed in any of the three motions.

Third rule: From an imperfect consonance to a perfect consonance one must proceed in contrary or oblique motion.

Fourth rule: From one imperfect consonance to another imperfect consonance one may proceed in any of the three motions.

After this rather challenging summary of the mathematical theory of music, presumably indicating what the prospective learner is expected to know, Part II begins the dialogue presenting counterpoint itself. Where Morley's Philomathes came not knowing a note, Joseph has already learned quite a lot of ancient theory. Indeed, Aloysius warns him about the "heavy task" of taking up a life's work in music: "You must try to remember whether even in childhood you felt a strong natural inclination to this art and whether you were deeply moved by the beauty of concords." Joseph responds: "Yes, most deeply. Even before I could reason, I was overcome by the force of this strange enthusiasm and I turned all my thoughts and feelings to music. And now the burning desire to understand it possesses me, drives me almost against my will, and day and night lovely melodies seem to sound around me."[40] We are far from the world of Morley's socially anxious gentlemen; Joseph is born to music and dreams only of pursuing it. Apart from that desire, he seems neither socially nor economically ambitious; Aloysius warns him not to expect to gain riches and possessions from music, from which we infer that Joseph probably has neither.

Further, the climb of Parnassus seems also to be a voyage into the past. Palestrina was born two centuries before this book, which nevertheless is addressed to contemporary composers, tasked to master a musical idiom of the past before even beginning to write in the style of their own time. By comparison, Zarlino and Morley based their precepts on the practice of their own teachers, Byrd and Willaert, though Zarlino also looked further back to Josquin. Joseph, however, never questions the need to make this journey into the more distant past, as if the knowledge to be gained was so fundamental that only an ancient master could convey it. By choosing to write in Latin, Fux further stationed his work as part of an ancient and enduring tradition, even as he presented it to modern students.[41]

Having defined counterpoint to be "a composition which is written strictly according to technical rules," Fux then introduced the concept of *species*, another helpful innovation. Though Zarlino and earlier writers had used "species" to describe each different kind of

Figure 9.3
Fux's first example of species one counterpoint, showing the numbering of intervals between the two voices (here the given cantus firmus is the lower line, the composed counterpoint the upper); the measures are numbered sequentially above the top staff (♪ sound example 9.1).

musical interval (such as thirds or fifths), Fux used this term to describe each of a series of five progressively more complicated forms of counterpoint. From these contrapuntal species, Fux considered two-part composition to be a single genus, seeing three- and four-part compositions as separate genera. This terminology harks back to Aristotelian teachings about species as the fundamental kinds of beings but also parallels contemporary initiatives to classify minerals, plants, and animals, such as Carl Linnaeus's *Systema Naturae* (1735), published ten years after Fux.[42]

Fux's first species is literally note against note: each note of the given cantus firmus (all of equal length) is matched with one of the counterpoint. Aloysius provides a simple cantus firmus (in the first Gregorian mode), gives a few words of advice (begin and end with perfect consonances, use more imperfect consonances than perfect), and asks Joseph to write a counterpoint in species one. The rules are so clear and simple that he can readily do so; Joseph writes numbers to indicate what each interval is so that he can show that his solution "was not by accident but by design," a product of the strict rules rather than inspiration (figure 9.3; ♪ sound example 9.1). With such clarity, even a beginner can write a correct counterpoint. The book's open and inviting approach makes it an ideal text for self-study. As such, it remains in use even today; even students who initially consider composing music to require unaccountable intuition or genius are startled to find that they can write harmonious counterpoint just by following a few simple rules. Among my students, I have noted that sometimes those who begin as the most adamant advocates of arbitrary genius, and the most skeptical of the value of any analysis of music, ironically become the most zealous and rigorous enforcers of Fux's rules when they start to do their own counterpoint exercises.

As his title promises, Fux provides carefully graduated steps from one species to another, again reflecting post-Cartesian ideas of clarity and distinctness as touchstones of scientific knowledge.[43] This especially characterizes the all-important first steps of this study. Species one had involved only consonant intervals; in species two, Fux allows two notes in the composed voice against one of the cantus firmus (figure 9.4a, ♪ sound example 9.2a). Accordingly, the first of those two notes is consonant (the one on the downbeat), but

Figure 9.4

(a) Fux's example of species two, in which two notes of the counterpoint appear against each note of the cantus firmus. Note that in each case, the downbeat must be consonant, while the other note can be either consonant or dissonant (♪ sound example 9.2a). In measures 9 and 10, the master marks with a cross Joseph's violations of the rule forbidding parallel fifths, which are not "hidden" by the skip of a third between them. (b) Fux allows the dissonant fourth F to appear as a passing tone between the consonances G and E, a third apart—a technique he calls *diminution* (♪ sound example 9.2b).

now the second could be dissonant if it is a *passing tone*, filling out the space between notes a third apart (which he calls *diminution*, as in figure 9.4b; ♪ sound example 9.2b): "the middle note may be dissonant because all three of them move stepwise."[44] Thus, in species two Fux begins to allow the student to practice the control of dissonance by imposing just the limitation that Palestrina had implicitly obeyed.

The succeeding species of counterpoint allow further rhythmic activity while still maintaining the control of dissonance under the same basic rules. In species three, four quarter notes accompany every whole note of the cantus firmus, the quarter notes thus implicitly required to be alternately consonant and dissonant. In species four, Fux presents *syncopation*, in which the counterpoint is staggered rhythmically against the cantus firmus so that what had been a consonance on the upbeat turns into a dissonance on the downbeat (figure 9.5a; ♪ sound example 9.3a). This technique would later be called "suspension" because a sustained note becomes dissonant by being suspended over motion in another voice. Fux notes that, compared to the "nonessential" passing dissonances of species two and three—dissonances that are basically stepwise transitions between adjacent consonances—in contrast, suspensions are dissonant "functionally, and on the downbeat; and since they cannot please by themselves, being offensive to the ear, they must get

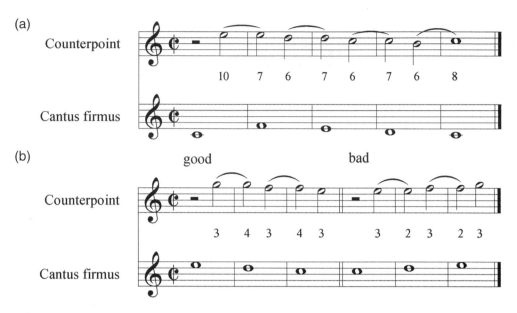

Figure 9.5
(a) Fux's example of species three (suspensions), in which the counterpoint is staggered rhythmically against the cantus firmus, so that (in measure 2) a dissonant seventh on the downbeat is resolved downward to a sixth (♪ sound example 9.3a). (b) Fux's examples contrasting the "good" resolution of dissonance downward against the "bad" resolution upward (♪ sound example 9.3b).

their euphony from the resolution into the following consonance."[45] Here Fux is dealing with one of the most important expressive and rhythmic devices of sixteenth-century counterpoint; Byrd, for one, used suspension beautifully (see figure 8.3) to create successive waves of expression.

At this point, Fux uses what he calls the "functional" quality of these dissonances to discuss the larger question of their resolution. He observes that held notes in these suspensions, "as it were, bound with fetters, are nothing but retardations of the notes following, and thereafter proceed as if brought from servitude into freedom. On this account dissonances should always resolve descending stepwise to the next consonances."[46] When Joseph asks whether they could also resolve by moving stepwise *upward*, Aloysius underlines this "problem which is harder to untangle than the Gordian knot" and states that "as your teacher I advise you to resolve all dissonances down to the next consonance" (figure 9.5b; ♪ sound example 9.3b), which later commentators glossed as meaning a "natural law of gravity" in the realm of music.[47] Though Fux does not make this connection explicit, he here seems to stand as a kind of musical Newton, proposing a law of universal downward resolution that Palestrina implicitly understood and that likewise should inform the practice of later composers who seek to learn the hidden laws of music.

Fux calls species five "florid counterpoint," which combines all the devices of the previous four species to produce the varied melodic lines of actual compositions, a garden "full of flowers." Thus, having broken down such lines into simpler constituents, Fux now shows how they can be combined and used together, by analogy with mathematics: "Just as we use all the other common species of arithmetic—counting, addition, multiplication, and subtractions—in division, so this species is nothing but a recapitulation and combination of all the preceding ones."[48]

After presenting all five species in two voices, Fux goes on to the next genus, three-part composition, which he considers "most perfect of all" because it includes all three notes of a "complete harmonic triad without adding another voice."[49] In the course of extending his rules to three and four voices, Aloysius notes that the nature of the intervals (and hence their function under the rules) "is to be determined by reckoning from the bass regardless of what may occur between the inner voices," assuming that those voices avoid fundamental mistakes such as parallel fifths or octaves.[50] Thus, the voices do not have equal function; they depend on the lowest sounding voice, the bass, in ways that go beyond the cantus firmus on which Fux initially taught his rules.

Though Fux does not comment further on this fundamental inequality between the voices, he does note that "the closer the parts are led together the more perfect the sound will be, for a power compressed will become stronger," as if in many-voiced composition the separate parts now cohere into an almost physical form, compressed and stronger than each of them apart from the others. Thus, to avoid dissonance and obey these rules, the forty voices of Striggio's motet cohered so strongly that he summarized them by a continuo, a single bass line with some numbers indicating the harmony above it. Here we stand at the verge of harmony as a separate science from counterpoint. Where counterpoint treated its voices as equal constituents, coordinated by rules, the new science of harmony raised its view to the unity formed by those separate voices.

10 In Bach's Hands

J. S. Bach's compositional practice combined contrapuntal mastery with the exploration of a new world of harmony that lay beyond Fux's more restricted teachings. Looking back from the mid-twentieth century, Anton Webern saw Bach as the discoverer of every aspect of musical thought that mattered to him, for "*everything happens in Bach*: the development of cyclic forms, the conquest of the tonal field, and, with it all, staggering polyphonic thought!"[1] Bach consciously projected his understanding of musical science in his works, which he understood as teachings and exemplars of knowledge as well as works of art. In so doing, he brought together the older traditions of counterpoint with the expressive, dramatic art of the second practice. Besides his consummate skill with the polyphonic forms of his time, Bach also pioneered a new kind of virtual polyphony in a single voice. As we shall see, his keyboard works depended on a new "digital" technology: the equalization of the thumb with the other fingers, essential for Bach's complex contrapuntal structures. Using this technology, he created exemplary fugues that served as idealized models of mental function, virtual minds that conversed or argued with themselves.

As Christoph Wolff has argued, Bach thought of himself as a learned musician, a scientist no less than master artist.[2] His contemporaries compared Bach to Newton, each having made fundamental advances in their respective domains of science.[3] As a member of the Correspondierte Societät der musicalischen Wissenschaften (Corresponding Society of the Musical Sciences), Bach was part of an elite association of intellectually minded musicians founded in 1738 by his student Lorenz Christoph Mizler. Bach's portrait shows him holding the canon (BWV 1076) he wrote for his entry to that society (figure 10.1; ♪ sound example 10.1), a work demonstrating his total mastery of Fuxian counterpoint; this picture demonstrates that contrapuntal ingenuity was an essential part of the way Bach wished to be seen. Indeed, the intercrossing voices of this canon build up a primordial harmonic progression (an authentic cadence, dominant to tonic), endlessly recirculating, as if demonstrating that harmony should be understood as resultant from counterpoint, rather than vice versa.[4]

The study of counterpoint was a central aspect of musical science as Mizler and Bach understood it. In fact, Mizler was responsible for the first translation of Fux, into German

Figure 10.1
Bach's portrait by Elias Haussmann (1748); the detail shows the triple canon in six voices that Bach holds
(♪ sound example 10.1).

(1742), prepared "under the very eyes of Bach, as it were."[5] Mizler's translation took care to include Fux's mathematical Book I, which subsequent translations into other languages tended to omit; in his very first annotations, Mizler mentioned Kepler's planetary harmonies and "the immortal Newton," giving further evidence that Bach himself was aware of their work.[6] Yet though he esteemed Fux and shared his devotion to musical science, in his own teaching Bach did not use Fux's *Gradus ad Parnassum*, even though he probably supervised and advised Mizler's translation. According to his son Carl Philipp Emanuel, because Bach

himself had composed the most instructive pieces for the clavier, he brought up his pupils on them. In composition he started his pupils right in with what was practical, and omitted all the *dry species* of counterpoint that are given in Fux and others. His pupils had to begin their studies by learning pure four-part thoroughbass [*basso continuo*]. From this he went to chorales; first he added the basses to them himself, and they had to invent the alto and tenor. Then he taught them to devise the basses themselves. He particularly insisted on the writing out of the thoroughbass in [four] parts.[7]

To complete these assignments, Bach's students implicitly had to apply Fux's rules (which Bach seems to have assumed they knew) in the context of a particularly important contemporary musical texture—the four-part chorale—rather than writing imitations of Palestrina. In that sense, Bach chose the "practical" over the antiquarian, but the rules remained the same; by writing out the chorales, his students also prepared to extemporize continuo to accompany vocal or instrumental ensembles, an essential skill for keyboard players at the time. These were very much "hands-on" exercises: his last student, Johann Christian Kittel, recalled that, at their gatherings,

one of his most capable pupils always had to accompany on the harpsichord. It will easily be guessed that no one dared to put forward a meager thoroughbass accompaniment. Nevertheless, one always had to be prepared to have Bach's hands and fingers intervene among the hands and fingers of the player and, without getting the way of the latter, furnish the accompaniment with masses of harmonies that made an even greater impression than the unsuspected close proximity of the teacher.[8]

We will shortly return to these intertwining hands and fingers.

Preferring this direct practical approach, Bach chose not to write a treatise of his own, leaving his students to undertake that labor, including Carl Philipp Emanuel, Mizler, Johann Philipp Kirnberger, and Friedrich Marpurg, all of whom wrote important theoretical works. As Carl Philipp Emanuel emphasized, many of Bach's own compositions explicitly claim to teach as well as to delight or refresh. For his family and students, though, he did write out summary notes to guide their beginning steps in thoroughbass, showing how important he considered it.[9]

In a way, Bach's teaching strategy guided his students more closely than the initial stages of Fux's text. By having to write an alto part (say) between a given chorale melody and Bach's own bass line, the student is more tightly constrained by the *musical* context than by Fux's two-part exercises. Such learners would have to track sensitively the voices above and below, as if Bach were guiding them not through rules alone but by the musical demands of creating a singable line that would fit the given melodies around it (of course avoiding the parallel motions that Fux's rules forbid), then comparing their solutions to what Bach himself had done in his own settings of that chorale. His compositional teaching, like his instrumental instruction, put his fingers right in the midst of his pupil's.

Then too, the students (like everyone else in their milieu) would have been intimately familiar with those hymns, having heard and sung them all their lives, so that they would be sustained and guided by their close acquaintance with that music as they practiced writing inner voices for those well-known chorales. Similarly, when he included such chorales in the *St. Matthew Passion*, Bach expected that his congregation would thereby be even more deeply drawn into his monumental work through their shared experience of singing those very chorales in the dramatic context he provided. Far from being merely functional church music, Bach's chorales are masterpieces in themselves; indeed, much of the development of music in the following centuries could be considered applications of the chorale to more or less extended dimensions, for instance in Beethoven's slow movements or Schumann's character pieces.

Even today, when fewer and fewer students grow up knowing the Protestant hymnody so familiar from Bach's time until the middle of the twentieth century, following his path of chorale writing is still the center of harmonic instruction in many colleges and universities. In that sense, we are all Bach's students now. No pedagogical method has superseded the powerful experience of trying to write an inner voice for a chorale and comparing it to what Bach did, seeing how he confronted the same problems and weighed the possible solutions.

Bach carried this same approach from the most fundamental elements of chorale setting to the most elaborate forms, such as canon, which he (like Fux before him) considered a summit of musical science. Thus, to demonstrate his worthiness to join Mizler's Society, Bach wrote *Some Canonic Variations on the Christmas Hymn "Vom Himmel hoch da komm ich her"* (*Einige canonisches Veraenderungen über das Weynacht-Lied: Vom Himmel hoch, da komm ich her*, BWV 769, 1747). Despite its modest title, this organ work contains some of Bach's most challenging counterpoint, whose modern stance and intricacy moved Igor Stravinsky in 1956 to make an orchestral arrangement that would popularize this little-known masterpiece. In it, Bach seems to underline the connection between the simple Christmas hymn and the triple canon in his portrait (figure 10.1), both sharing the same melodic figures. Luther himself was credited with this tune, whose alternately falling and rising scale figures seems to depict the savior's descent from heaven and return thither, illustrating its text: "From heaven on high I came here."

Having advised his students to begin with a hymn tune and write chorale settings of it, with this work Bach shows how he would carry that assignment even further. His first two variations present canons on themes related to the hymn (at the octave and fifth, respectively, above the original voice), while the pedal plays the hymn tune in much longer notes (augmentation). The third variation demonstrates the inclusion of a rhapsodic, freely composed alto voice that intertwines with a strict canon between two other voices; above them floats the hymn tune in the soprano (figure 10.2). Stravinsky's orchestration assigned this free alto voice to the English horn (cor anglais), echoing Bach's soloistic use of the similarly dark timbre of the oboe d'amore in his cantatas and passions. Though Vincenzo Galilei had argued that modern counterpoint abandoned the expressive power of ancient melody, in this variation (and many of his other works) Bach demonstrates how the expressive flexibility of one voice can be combined with strict counterpoint, indicating the synthesis of the two streams of music we have been tracking separately, thus uniting first practice and second, ancient and modern, dispassion and expression.

The fourth variation takes a further step in contrapuntal ingenuity, a canon in augmentation: the second voice plays the same material twice as slowly as the first, against a free

Figure 10.2
Variation III of Bach's canonic variations on "*Vom Himmel hoch*" (♪ sound example 10.2): the pedal and left hand play a canon in eighth notes at the seventh based on the hymn tune, which appears in half notes in the soprano. The alto voice (marked *cantabile*, singable) freely intertwines between these other voices.

third voice, the hymn tune still in the pedal part. The fifth and final variation contains successive canons at the sixth, third, second, and ninth, in which the second voice answers the first played *backward* (retrograde), over a walking bass in the pedal. In the final three measures, Bach combines all four phrases of the hymn in very close overlap (*alla stretta*), almost simultaneously, some of the voices playing twice as fast as others (diminution), as if his climax compressed the whole hymn into three explosive measures (figure 10.3; ♪ sound example 10.3). The very last measure even includes a hidden signature, B–A–C–H (German notation for B♭–A–C–B♮) as well as the enormous stretch of a tenth in the right hand (A–A–C), as if marking the imprint of Bach's own huge hands, said to span a twelfth.[10] Though presented in all humility, so superhuman is this contrapuntal accomplishment that one might wonder whether its author himself came down from heaven on high. Writing in 1754 about these canonic variations, Johann Michael Schmidt underlined the comparison with mathematical science: "I cannot persuade myself that the most difficult demonstrations of geometry requires much deeper and more extensive reflection than this labor must have demanded."[11]

Besides their brilliance in the science of counterpoint, these variations also demonstrate a technological advance Bach made in keyboard playing that revolutionized its polyphonic possibilities: a new kind of fingering that made the thumb the equal of the other fingers. Carl Philipp Emanuel recalled his father telling him "that in his youth he used to hear great men who employed their thumbs only when large stretches made it necessary. Because he lived at a time when a gradual but striking change in musical taste was taking place, he was obliged to devise a far more comprehensive fingering and especially to enlarge the role of the thumbs and use them as nature intended," so that "they rose from their former uselessness to the rank of principal finger."[12] Bach's innovation in fingering quickly became so standard that today we tend to forget its origins and significance, yet it affected keyboard

Figure 10.3
The concluding three measures of Variation V of Bach's *Canonic Variations on the Christmas Hymn "Vom Himmel hoch"* (♪ sound example 10.3); in this compressed passage, all four phrases of the hymn tune are played in close overlapping (stretto), some in eighth notes (alto and bass), others in sixteenth notes (soprano and tenor). Note also the huge stretch in the right hand in the third beat of measure 55, spanning a tenth (A–A–C) and the presence of Bach's musical signature B♭–A–C–B♮ (= H in German notation) in the final measure.

playing so deeply that it may be more apt to call it a technological (rather than merely technical) advance, because it fundamentally reconfigured the use of the human hand.

Many fingered scores from the fifteenth and sixteenth century specified that the right thumb was not to be used at all, using the left thumb only for wide stretches, as was the practice of Christian Erbach (figure 10.4; ♪ sound examples 10.4a,b), described in 1614 as "the best organist and composer in Germany."[13] Indeed, even the numbering of the fingers reflected this practice, some sixteenth-century scores using "0" to denote the thumb as "null," reserving "1 2 3 4" for the real fingers.[14] Contemporary graphic images showed players holding their wrists so low that their thumbs are below the keyboard (figure 10.5).[15] Even in 1753, Carl Philipp Emanuel described keyboard players "who do not use the thumb [and] let it hang to keep it out of the way," which he thought stiff and uncomfortable: "Can anything be well executed this way?"[16]

Avoiding the thumb involved crossing the index, middle, and ring fingers (2 3 4) over each other, creating a very different articulation from the smooth connection between the

Figure 10.4
The beginning of Christian Erbach's *Ricercar* (ca. 1625), in modern notation; the thumb is used only by the left hand, while the right hand plays scales using mostly the index and middle fingers (♪ sound examples 10.4a,b, played in the original and modern fingerings, respectively).

Figure 10.5
Anonymous woodcut from the title page of Nikolaus Ammerbach's *Orgel oder Instrument Tabulatur* (1571),
a work J. S. Bach owned, showing the player's right thumb held below the keyboard; an assistant operates the
bellows of his small organ.

notes (legato) that later became customary as a result of passing the thumb *under* the fingers; since the later twentieth century, some historically informed players try to use contemporary fingerings as part of their endeavor to render the music as closely as possible to the performance practice of the time.[17]

In 1717, François Couperin's *L'art de toucher le clavecin* (*The Art of Playing the Keyboard*) used the right thumb to begin scale passages, but then reverted to crossing the middle fingers over each other without using the thumb (figure 10.6a; ♪ sound example 10.5a). He compared this and similar fingerings in the "modern style [*façon moderne*]" to the "old style [*manière ancienne*]" that avoided the thumb, noting that because the old style "had no connection [*liaison*]" between notes, it had fallen out of fashion because "few people in Paris remain infatuated with old maxims, Paris being the center of what is good."[18] Nonetheless, Couperin tended to avoid using the right-hand thumb, using the other fingers to create small, articulated groupings, often involving a finger sliding from one key to another or substitutions of different fingers on the same note (figure 10.6b; ♪ sound example 10.5b); he clearly preferred the elegance, clarity, and articulate expressivity those fingerings promote.

Figure 10.6
(a) Couperin's fingering for an octave scale, from *L'art de toucher le clavecin* (1717), in which the right hand begins on the thumb but then uses the alternation of third (middle) and fourth (ring) fingers (1 2 3 4 3 4 3 4 5) ascending; the descending pattern is different (♪ sound example 10.5a); compare this fingering with the modern 1 2 3 1 2 3 4 5. (b) Couperin's fingering for his pieces "Les sylvains," in which the expressive grouping of two notes is underlined by the use of adjacent fingers, the third finger sliding between groups to make a slight articulation (♪ sound example 10.5b). Here, using the thumb would have created too much connection between groups.

Bach's first biographer, Johann Nikolaus Forkel, noted that he was "acquainted with Couperin's works and esteemed them."[19] Yet at the time Couperin published his fingering system, Bach was "above thirty years old and had long made use of his [own] manner of fingering," according to Forkel; though these two systems had in common "the more frequent use of the thumb," only Bach made the thumb "a principal finger because it is absolutely impossible to do without it in what are called the difficult keys," those with many sharps or flats, whereas Couperin "neither had such a variety of passages, nor composed and played in such difficult keys as Bach, and consequently had not such urgent occasion for it [the thumb]."[20] Wolff considers the "revolutionary impact" of Bach's and Couperin's use of the thumb a case of independent invention, mentioning another such case, the famous priority dispute between Newton and Leibniz about the invention of calculus.[21]

We can gain a sense of the development of Bach's fingerings in his *Applicatio* (BWV 994), written in a notebook for his eldest son, Wilhelm Friedemann (figure 10.7, ♪ sound example 10.6). Though written about 1720 (hence long after Bach himself adopted his "new" fingerings), he teaches his ten-year-old son many aspects of the older practice; the right hand begins by playing a scale with alternating inner fingers (3 4 3 4), but then uses the thumb before a leap (at the end of the bar).[22] Yet he fingers the left-hand scale (in measure 3) with alternating thumb and index fingers (2 1 2 1), indicating how natural he thought this thumb crossing (which nowadays would be considered rather awkward). Bach himself was left-handed, a trait that conferred "no small advantage on our instrument," according to Carl Philipp Emanuel, whose own left-handedness was so marked that (unlike

Figure 10.7
J. S. Bach's *Applicatio* (BWV 994), showing the composer's own fingerings, written to teach his eldest son, Wilhelm Friedemann. Note the use of finger crossing (3 4 3 4) in the right hand (measure 1) and thumb crossing (3 2 1 2 1 2 1) in the left (measure 3); the right hand uses thumb crossings in measure 7, when three voices are heard (♪ sound example 10.6).

his father) he avoided learning to play string instruments.[23] Arguably, the elder Bach's left-handedness led him to find comfortable this thumb crossing in the left hand, a practice he may then have applied to the right (as he does in measure 7). Having thus learned from his own left-handedness, Bach may have decided that he would use similar principles even for the right-handed, such as his eldest son, Wilhelm Friedemann.

Bach's revolutionary "digital technology" emerged from the needs of polyphony, even more than from the need to play in keys remote from C. For instance, his *Applicatio*, written in the purest C major, uses the passing of the right thumb precisely in the passage that involves three voices, two of them in the right hand (measure 7). Thus, he found that enabling this additional voice in the same hand required the use of the thumb. Any page of Bach's complex counterpoint, in any key, will quickly prove the necessity of his new thumb technology, especially for his fugues, a complex polyphonic form of which he was considered by many contemporaries to be the supreme exponent.

In his general critique of contemporary counterpoint, Vincenzo Galilei had been particularly critical of the technique of imitation then called *fuga*, literally a "flight" of fancy that structured canonic imitation of a theme sung against itself, alternately appearing a fifth higher (as an "answer" to the original subject), then in its original form.[24] The overlaid statements of the theme against itself came to be called "entrances." These entrances were interspersed with "episodes," more freely written sections. Over time, fugue as a musical form acquired the reputation of being the most demanding form of counterpoint; Vincenzo attacked it as the epitome of the dispassionate contrapuntal style he wanted to replace with dramatic declamation. Yet Bach's fugues synthesized these two styles Vincenzo thought were so antithetic.

Bach solved these difficulties through cultivating canonic forms combined with freer composition. Reflecting on his experiences with Bach as a teacher, Kirnberger noted that "his method is the best, for he proceeds steadily, step by step, from the easiest to the most difficult, and as a result even the step to the fugue has only the difficulty of passing from one step to the next."[25] Bach was able to teach in this sequential way because he himself had cultivated a systematic analysis of the contrapuntal possibilities that a given theme could allow: for instance, could it be played against itself in inversion? Or in diminution? Carl Philipp Emmanuel recalled that,

when he listened to a rich and many-voiced fugue, he could soon say, after the first entries of the subjects, what contrapuntal devices it would be possible to apply, and which of them the composer by rights ought to apply, and on such occasions, when I was standing next to him, and he had voiced his surmises to me, he would joyfully nudge me when his expectations were fulfilled.[26]

The elder Bach's insight and experience of fugue writing were as keen as his generous joy when someone else was also able to realize the full range of contrapuntal possibilities.

Consider Bach's *Ricercar à 6*, a six-part fugue he wrote at the challenge of the Prussian king Frederick II (who employed Carl Philipp Emanuel as court musician).[27] First published after Bach's death in open score (six staves, one for each voice), the better to display the independent details of each separate voice, it is surely no accident that his fugue can be played on a single keyboard by one player (♪ sound example 10.7). In a way, this work is comparable to Tallis's forty-part motet as polyphony operating at its limits, not through the sheer number of voices—others had written six- or eight-part fugues—but by having six independent voices played by ten fingers. Anyone who plays it realizes the absolute necessity of Bach's innovative use of the thumb as "principal finger"; even so, this fugue tests the most flexible hands.

Despite the digital gymnastics involved, Bach's six-part writing gives a sense of total freedom and ease; moving within the tightest contrapuntal demands, its voices seem expressive and unconstrained. Indeed, the voices seem to inspire each other to greater heights of intensity, contradicting Vincenzo's opposition of contrapuntal rigor to the emotional expressivity Bach also prized. According to his student Friedrich Marpurg, Bach "pronounced the works of an old and hardworking contrapuntist *dry and wooden*, and certain fugues by a more modern and no less great contrapuntist."[28] Bach found these fugues "*pedantic*; the first because the composer stuck continuously to his principal subject, without any change; and the second because, at least in the fugues under discussion, he had not shown enough fire to reanimate the theme by interludes," meaning the more freely composed episodes between entrances of the theme. To contemplate Bach's own fire, we will consider another of his personal innovations in fugue.

The explicit challenge of the *Ricercar à 6* involved the interplay of six independent voices under two hands, but Bach's polyphonic vision reached further still to the seemingly paradoxical possibility of polyphony *within a single voice*. He devoted special attention to what one might call virtual or implicit polyphony (some writers call it "pseudo-polyphony") possible even in the context of a single melodic line. The clearest instances of this can be found in Bach's works for solo violin or cello, in which the solitary string instrument might be taken as the purest possible exponent of melody. For instance, the first movement (Prélude) of Bach's Fifth Suite for solo cello begins with a searching exploration of the implications of its opening octave C, unfolded in what seems an almost improvisatory style, though a deep structure underlies his successive points of arrival. Here, occasional chords involving several of the cello's four strings punctuate a meditative melodic line (figure 10.8; ♪ sound example 10.8).

Yet the seemingly free exploration of this opening (once it has finally and deliberately reached the dominant, G) turns out to be the prelude—literally the preparation—for an extended fugue in which the cello only very occasionally plays more than a single simultaneous note. How is this possible? The solo cello in fact plays all of the different voices itself not simultaneously but in succession, jumping between them, using suggestion to allow us to infer that there are, in fact, several voices at work, not just one (figure 10.9; ♪ sound example 10.9). Indeed, the very opening of the fugue's subject (measures 27–33) implies two voices in dialogue. These two voices are separated by leaps, one starting when the

Figure 10.8
The opening of the Prélude of Bach's Fifth Suite for cello solo (♪ sound example 10.8).

other falls silent, as if answering each other, essentially an extended use of the technique of hocket. When the theme returns in yet another imagined "voice" at the end of measure 35, Bach has already begun to elaborate the theme, adding additional ornaments and hence fire to its new appearance.

Bach thus not merely gestures to counterpoint but writes an extended fugue with many increasingly ornamented episodes, perhaps to help camouflage the artifice by which he suggests additional voices within a single melodic line. Indeed, this fugue overall has about the same number of measures as the *Ricercar à 6*. The cello fugue also moves toward a climactic final entrance, in which the instrument leaps down to its lowest register (which Bach reserves for this moment) to give additional weight to this peroration (figure 10.10; ♪ sound example 10.10). Though throughout he has used elements of the prelude's linear texture to elaborate his fugal lines, Bach seems to reintegrate these improvisatory elements even more closely in his final passage, thus synthesizing solo improvisation with the rigor of counterpoint. Where the introductory prelude had been in 4/4 time, as if an exalted march, the fugue is in 3/8, a dance-like triple meter that gives new life and animation to the stately meditations that preceded it.[29]

Figure 10.9
The beginning of the fugue in Bach's Fifth Suite for solo cello (♪ sound example 10.9) as virtual polyphony.
Note that the initial notes G–A♭ (measure 27–28) in one "voice" are answered by C–D–E♭ in the other "voice"
(indicated by dotted parentheses), and so forth. There are other entrances at measures 35 and 48, separated by
episodes.

It is as if this fugue, in its triumphant conclusion, returns carrying the rhapsodic prelude that had given it birth. Bach thus draws together the seemingly divergent strands of
monophony and polyphony by showing that a single melodic line could present itself as
suggesting many voices in one. This solo fugue operates through artful silences and suggestions, allowing the mind to leap back and forth between voices that otherwise could
be understood as a single melody. If this leaping monophony could create the illusion of
polyphony, perhaps even simultaneously sounding voices could likewise be understood
one at a time by a mind that leaped between them. In the process, Bach demonstrates the
integral connection between utmost expressivity and contrapuntal rigor, making his polyphony inextricable from expression, from the free flight of the mind itself.

Further, Bach helps us reconsider the mind's flight and its body via the intermediacy
of the hands. In his *Ethics* (1677), Benedict de Spinoza had argued that, "to the extent
that some body is more capable than others of doing several things at the same time, or
of being acted on at the same time, to that extent its mind is more capable than others of
perceiving several things at the same time."[30] Spinoza's insight implies that the mind's
ability to process polyphony should really be attributed to the body, particularly the hands,
rather than to disembodied mental faculties. If so, Bach's "thumb revolution" opened a new

Figure 10.10
The conclusion to the fugue in Bach's Fifth Suite for solo cello (♪ sound example 10.10).

mental world of polyphony because the newly empowered hands enabled a new degree of complexity for a solo keyboard fugue.[31] Still, Bach made fun of people whose ability to compose came solely from the way they navigated the keyboard, calling them "keyboard knights [*Clavier Ritter*]," implying that such purely keyboard-driven playing was comically limited.[32] Though he himself was said to have composed without using an instrument, the figurations of his more improvisatory-sounding works seem to owe something to his actual improvisations at the keyboard. Nevertheless, even if not composed at a keyboard, everything he wrote for it reflects the capabilities of his virtual hands.

Many contemporaries judged Bach to be the greatest exponent of the fugue, considered as a rhetorical and mental construct of particular importance. Already by Bach's time, fugue had a special status even among contrapuntal forms. A long tradition considered a fugue to be a kind of "free oration," whose structure reflected the felt dynamics of human reason and persuasion, compared to the canon as a "strict oration" that deployed the figures of musical logic more austerely.[33] Thus, both fugue and oration had similar sequences of sections, leading from an initial introduction of its topic (the entrances), to their exam-ination and development (episodes), ending with a final peroration that would reiterate and drive home their point. The contemporary theorist Johann Mattheson thought that the theme of a fugue corresponded to "what for the orator is the theme or subject."[34] Returning to the imagery of personhood we have been tracing since the Middle Ages, Forkel noted that Bach "considered his parts as if they were persons who conversed together like a se-lect company. If there were three, each could sometimes be silent and listen to the others till it again had something to the purpose to say."[35] Even as late as 1870, Richard Wagner described a theme as "a person and not a discourse."[36]

Some eighteenth-century writers compared the rhetorical structure of fugue to a "strug-gle" (*Streit*) between adversaries, emphasizing the element of combat or contest, while others depicted the process of persuasion as a more peaceful unfolding of argument and example. Thus, for Mattheson "each fugue has, so to speak, two principal combatants who have to settle the issue with one another," namely the fugue's theme (*dux* or "leader") and its countersubject (*comes*, "follower").[37] For his older contemporary Johann Chris-toph Schmidt, the follower was not an adversary but actually offered an explanation of its leader.[38]

After introducing his conflicting protagonists, Mattheson viewed the ensuing fugal dis-course as enabling what he called *confutatio*, "a resolution of the objections" in rhetoric, here meaning the conflicts between leader and follower that "may be expressed in music either through suspensions or also through the introduction of strange-seeming [*fremds-cheinender*] passages. For it is just by means of these elements of opposition, provided that they are rendered prominent, that the delight of the ear is strengthened and every-thing in the nature of dissonances and syncopations which may strike the ear is settled and resolved."[39] In this process, Forkel also emphasized the importance of those "strange" passages—literally "foreign" (*fremd*)—that characterized Bach's style in the minds of his contemporaries. Already in 1625 Francis Bacon had noted that "There is no Excellent *Beauty*, that hath not some Strangenesse in the Proportion," thus indicating a significant shift between older notions of beauty as symmetry and the newer taste for a style that re-quired "strangeness" to reach resolution—"Baroque" style, to use a later term describing a distorted yet wonderful pearl.[40] For Forkel and Mattheson, such beautiful strangeness re-sulted from Bach's use of the contrapuntal artifices of inversion or augmentation to achieve dramatic intensification.

All these contrapuntal techniques culminate in a concluding peroration, which Forkel called "the ultimate, strongest repetition of such phrases as constitute, as it were, a conse-quence of the preceding proofs, refutations, dissections, and confirmations … for the pur-pose of completely driving home for the listeners the emotion aimed at in the piece one last time."[41] Mattheson noted that this "epilogue" calls for "a fundamental note" like a pedal point, the sustained sounding of the tonic, such as Bach used in figure 10.3 to underline his final spectacular measures, "or a more intense accompaniment," such as underwrote his solo cello fugue (figure 10.10; ♪ sound example 10.10).[42] These rhetorical and musical techniques bring their arguments to a conclusion as decisive as the QED that ends a mathematical demonstration.

In view of these means and ends, these commentators considered a fugue (like an oration) to be a heightened exercise of human reason, feeling, and persuasion that required and exemplified those qualities. If indeed its form instantiates intelligence functioning in its true capacity of discourse and argument, a Bach fugue represents a kind of virtual mind, human in structure and feeling yet suggesting an almost divine capacity.

III POLYPHONIC HORIZONS

11 Polyphony Extended

In the centuries following Bach, the marriage between polyphony and homophony generated many new hybrid works with varying degrees of independence between voices. Their ever-increasing dissonance reflected the fundamental tensions, even instability, of the project of music as expression. Examples from Ludwig van Beethoven, Johannes Brahms, and Arnold Schoenberg illustrate a growing interest in the possibilities of polyphonic complexity and its artistic effects, including new ranges of orchestral timbre. These explorations intersected with the physical and physiological studies of Hermann von Helmholtz, who explained dissonance in terms of the degree of clashing between overtones. He also described different timbres as produced by different strengths of overtones that would "color" a sound. The resultant possibilities fueled explorations of orchestral color by Schoenberg and others, leading to further expansions of polyphonic possibilities: simultaneous different rhythms, even tempos, such as Charles Ives, Karlheinz Stockhausen, György Ligeti, and Conlon Nancarrow employed.

Imagine a numerical index that measures the felt number of independent voices, an index that could vary considerably even within a given composition as its various constituent voices sound more or less independent to a listener. Already in the monody used in early opera, that index might have varied between one and two, as the ear felt the accompaniment to be more or less an audibly independent or divergent voice from the sung melody.

As with Tallis's forty-part motet, similar factors affected the many instrumental lines in an orchestra: a hundred or more players would find themselves actually playing only a few different pitches at any given time, constrained by triadic harmony and the degree of dissonance their listeners could tolerate. For instance, in a random passage from Beethoven's Third Symphony (figure 11.1; ♪ sound example 11.1), many instruments play subsidiary lines that add texture but are not meant to be heard as significant (or autonomous) as the melody and bass, which are themselves bound together harmonically.

Arnold Schoenberg used the term *polyphonic homophony* to suggest this complex mixture of contrapuntal elements that could be interwoven within a larger texture that is basically homophonic, its many voices closely correlated and acting as one.[1]

Figure 11.1
Ludwig van Beethoven, Symphony no. 3 ("Eroica"), from the coda of the first movement, measures 655–663
(♪ sound example 11.1); though the full orchestra is playing, there are essentially only two distinct contrapuntal
voices (the melody in violas, celli, and basses, against a countermelody in the winds), plus textural and rhythmic
material in the brass and violins.

Schoenberg especially had in mind composers such as Brahms, who often used rather prominent contrapuntal lines to give richness and complexity to a simpler basic texture. One might find many passages, such as one from Brahms's Second Symphony (figure 11.2; ♪ sound example 11.2), that could illustrate this characteristic sound of so much orchestral music from the nineteenth century. Yet the fundamental homophony of this music is illustrated by a story about Brahms, who, when he studied a new score, would use two fingers to cover the middle voices in order to isolate the melody and bass, which form the underlying skeleton on which all the other voices would rest.[2] His gesture suggests that those middle voices were textural additions, not as essential as the treble and bass.

To create a more complex texture, Brahms often used polyrhythm, the simultaneous playing of several different rhythms. The simplest example would be one hand playing a rhythm of two beats per measure, the other three per measure ("two against three").[3] Even such a simple polyrhythm was avoided by Schubert in 1827, as figure 11.3 (♪ sound example 11.3) illustrates. Though Chopin also tended to avoid polyrhythms in his earlier works, in 1840 he published a study that presented a technique for playing them that has remained in use to this day. Box 11.1 explains this technique for the case of two against three; Chopin's étude teaches it in the harder case of three against four (figure 11.4; ♪ sound example 11.4). He asks the pianist to play first the rhythm of three in one hand alone, then four in the other, so that each hand establishes its own independent felt pattern. Only after nine measures does he allow both hands to play together. In this way, he avoids the labored (and insufficiently independent) effect that would have resulted had the player initially tried to force the two rhythms to "fit together." For Chopin and Brahms, the expressive tension between two simultaneous rhythms became an important compositional device, as it continued to be thereafter. When, for instance, in 1908 Stravinsky's first étude for piano (op. 7) included the layering of five against three against two, he introduced his rhythmic layers separately, giving the hands time to absorb the unfamiliar feeling of playing groups of five, creating the turbulent feeling of this piece (figure 11.5; ♪ sound example 11.5).

Box 11.1
Playing two against three

Pick a constant overall tempo, say sixty beats per minute, which is 60 on the metronome: one beat per second. Practice having the left hand (say) tap out two beats per second; then have the right hand tap three beats per second. Practice each hand separately so that it feels very comfortable by itself (the other hand might just play the larger beat, one per second). Then practice going back and forth between hands; the left hand might do ten repetitions of two beats per second, followed by ten repetitions of three beats per second in the right hand. When each feels very easy and you can shift readily from one hand to the other, try playing both simultaneously without trying to "make" them fit.

Figure 11.2
Johannes Brahms, Symphony no. 2 in D major, op. 73, first movement (measures 204–212) (♪ sound example 11.2). Note the main melody in the first violins, against the countermelody in the violas; meanwhile, the celli and basses play another countertheme, answered in inversion by the clarinets. The second violins add another answering motive, rhythmically displaced by accents.

Figure 11.3
The beginning of Franz Schubert's "Wasserfluth," from his song cycle *Winterreise* (1827), showing an 1895 edition that follows the orthography of the manuscript. Note that he lines up the last note of the triplet with the last sixteenth note (especially at the end of measure 3, showing that he wished there to be no difference between these rhythms as played here (♪ sound example 11.3).

Further using polyrhythms, Schoenberg gave an even larger scope to the polyphonic elements in such compositions as his Chamber Symphony, op. 9 (1907), in which the hearer is made more insistently aware of a greater number of independent voices than in a symphony by Brahms (figure 11.6; ♪ sound example 11.6).[4] Though in figure 11.6 there is still a key signature, the rhythm is oddly dislocated, the 3/4 waltz tussling with other figures in 2/2 rhythm. In the following years, this increasing degree of polyphony accompanied a breakdown of the tonal fabric, namely the control of a central key that would regulate the acceptable degree of dissonance. As Schoenberg consciously "liberated" dissonance from these centralized controls, his textures became harmonically much more complex than those of Brahms. Because expressive music constantly strove to be ever more expressive, sustaining any single mood (or tonality) would, after a time, seem boring and would call for a contrasting swing to another mood so as to rekindle the drama. Yet this endlessly restless quest to further intensify passion would itself become boring at a certain point: the project of expressive music seemed to overreach itself or at least to be essentially unstable. As dissonance became less controlled, Schoenberg and other like-minded composers felt the need for more polyphony to knit together musical textures that no longer were bound by the conventional harmonies of earlier times. We will later consider other works that moved toward the dispassionate stance of musical science.

To augment their polyphony, Schoenberg and many other contemporary composers (including Claude Debussy and Maurice Ravel) explored the myriad possibilities of orchestration that relied on the different timbres of instruments. Hermann von Helmholtz's treatise on physiological acoustics, *Die Lehre von den Tonempfindungen als physiologische Grundlage für die Theorie der Musik* (*On the Sensations of Tone as a Physiological Basis for the Theory of Music*, 1863), presented the fundamental understanding of sound and timbre that, in large part, still stands today. As his title indicated, Helmholtz oriented his

Figure 11.4
The first page of Frédéric Chopin's Nouvelle étude, no. 1 (1840), in which the right hand begins alone by playing steady triplets (three quarter notes in the time of the half note of the piece's tempo), then separately the left hand plays four eighth notes per half note. At measure 9, the two rhythms are combined, each hand having assimilated the independent feeling of its own rhythm (♪ sound example 11.4).

Fig 11.5
The beginning of Igor Stravinsky's Étude, op. 7, no. 1 (1908) for piano; the left hand begins playing quintuplets, joined then by the right hand playing first triplets, then duplets. In measures 5–6, all three rhythms are combined (♪ sound example 11.5).

treatise toward music and its historical development from earliest times until his time. He treated "the *Polyphonic* Music of the middle ages, with several parts, but without regard to any independent musical significance of the harmonies" as a crucial period between the initial "*Homophonic* or Unison Music of the ancients, to which also belongs the existing music of Oriental and Asiatic nations," and "*Harmonic* or *Modern* Music, characterized by the independent significance attributed to the harmonies as such."[5] Even apart from such considerations of polyphony or homophony, Helmholtz drew attention to the multiplicity inherent even in a single musical tone because of the overtones discovered by Descartes and Mersenne. Each tone produced by a voice or an instrument is not a pure pitch but is composed of a whole spectrum of overtones whose precise mixture the ear interprets as the timbre of that tone. Helmholtz devised new scientific instruments (such as the vibration microscope shown in figure 11.7a) that could observe the different strengths of the various overtones that identify a tone as produced by a string instrument (figures 11.7b,c). He compared this with the characteristic overtones for wind instruments and the various sung vowels. His treatment implied that even a single instrumental tone is a kind of polyphony whose perceived sound is the sum of its many constituent overtones and their relative strengths.

Helmholtz used these findings to set forward a new theory of consonance and dissonance that treated both as matters of degree, not distinguished absolutely as music theory

Figure 11.6
Arnold Schoenberg's Chamber Symphony no. 1, op. 9, at rehearsal number 38 (♪ sound example 11.6).

(a)

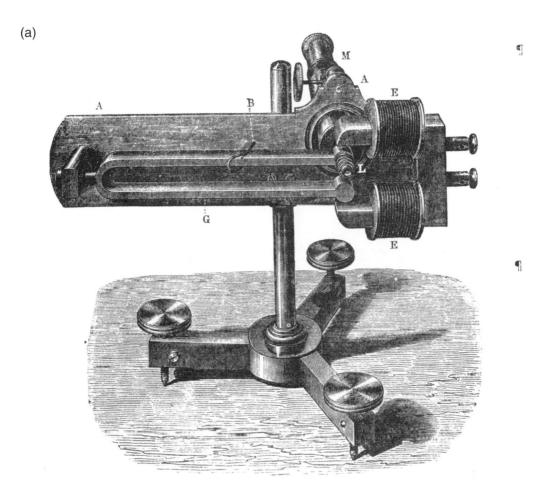

Figure 11.7
(a) Hermann von Helmholtz's vibration microscope, by which he observed the overtones of a violin string (to which a grain of starch was attached to make its vibrations more easily visible), from his *Tonempfindungen*. The tuning fork *G*, excited by an electromagnet *E*, is attached to a lens *L*, through which the violin string can be observed. (b) The visual patterns Helmholtz observed through his vibration microscope, in which *B* and *C* represent "the vibrational forms for the middle of a violin string, when the bow bites well, and the prime tone of the string is fully and powerfully produced." (c) Helmholtz's table of the relative intensity of different overtones of a violin string when plucked or struck at different points.

(b)

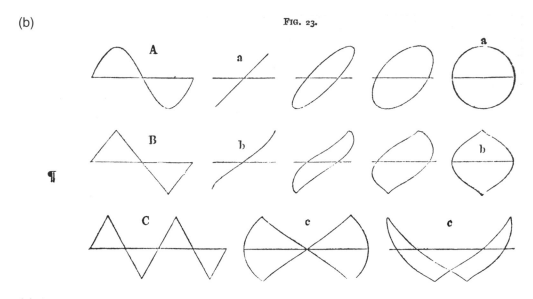

FIG. 23.

¶

(c)

Theoretical Intensity of the Partial Tones of Strings.

| Number of the Partial Tone | Excited by Plucking | Struck by a hammer which touches the string for | | | | Struck by a perfect hard Hammer |
| | | $\frac{3}{7}$ c'' | $\frac{3}{10}$ g' | $\frac{3}{14}$ $C_{,}-c'$ | $\frac{3}{20}$ | |
		Striking point at $\frac{1}{7}$ of the length of the string				
1	100	100	100	100	100	100
2	81·2	99·7	189·4	249	285·7	324·7
3	56·1	8·9	107·9	242·9	357·0	504·9
4	31·6	2·3	17·3	118·9	259·8	504·9
5	13	1·2	0	26·1	108·4	324·7
6	2·8	0·01	0·5	1·3	18·8	100·0
7	0	0	0	0	0	0

Figure 11.7 (continued)

up to the Renaissance had assumed. Given two different pitches, he calculated the relative degree of agreement of their overtones. For instance, if one violin plays a steady middle C and another violin produces a continuously sliding pitch that starts at that C and ends at the c an octave higher, at each point along that slide Helmholtz calculated the number of beats heard, meaning the pulses caused by the difference between the two tones (figure 11.8a). These more or less rapid beats, he argued, proportionally agitate the fibers within the inner ear located in the organ of Corti (figure 11.8b), which he thought was the seat of auditory sensation. That agitation gives rise to the relative sensation of "dissonance" or "consonance."

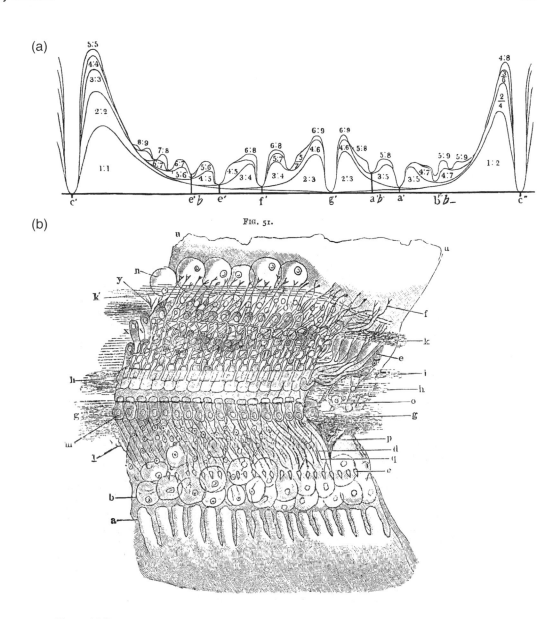

Figure 11.8
(a) Helmholtz's calculation in his *Tonempfindungen* of the relative intensity of beats heard between two pitches, one a middle C, the other continuously sliding from that C to the c an octave higher. Note the valleys at points of relative consonance (octave, fifth, fourth, third). (b) Helmholtz's illustration of the organ of Corti, whose microscopic structures in the inner ear respond to pitches.

Indeed, in figure 11.8a the "valleys" of relatively less beating correspond to the traditional ordering of consonance: the highest rate of beating corresponds to the dissonances of minor second and major seventh, the lowest to the consonant octave, fifth, fourth, and major third. If so, the sensations of dissonance and consonance result from a physical clash between different tones as they excite various fibers in the organ of Corti, which Helmholtz conceived as a kind of piano that would respond sympathetically to the incoming vibrations: an inner "instrument" responding to the instruments outside.

Going even further, Helmholtz emphasized that not even a pure tone (devoid of overtones) could really be considered a simple, continuous phenomenon. Though acoustics had historically grown from the study of continuously vibrating bodies such as strings, Helmholtz used the siren (a relatively recent addition to technology not yet in use as a musical instrument) as a more pure and reliable source of sound. Sirens produce sound by interrupting an incoming stream of air periodically, most simply by using a rotating disc with regular holes (figure 11.9a) that then generates a pitch that depends on the speed of rotation and the spacing of the holes. Thus, a pure pitch is simply a series of regularly timed puffs of air, as Oresme had speculated, not necessarily produced by a continuous process.[6] Helmholtz's polyphonic siren (figure 11.9b) allowed him to produce two simultaneous such pitches for experimental use; here, polyphony was simply the result of a single suitably contrived rotating disc, not of two different instruments.

(a)

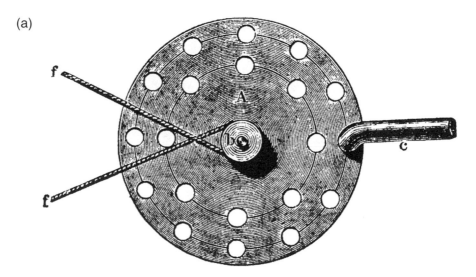

Figure 11.9
(a) Helmholtz's illustration of a simple Seebeck siren, the air stream (entering via tube *c*) regulated by a rotating disc with holes (*A*). (b) Helmholtz's own invention, the polyphonic siren, capable of sounding two pitches simultaneously.

(b)

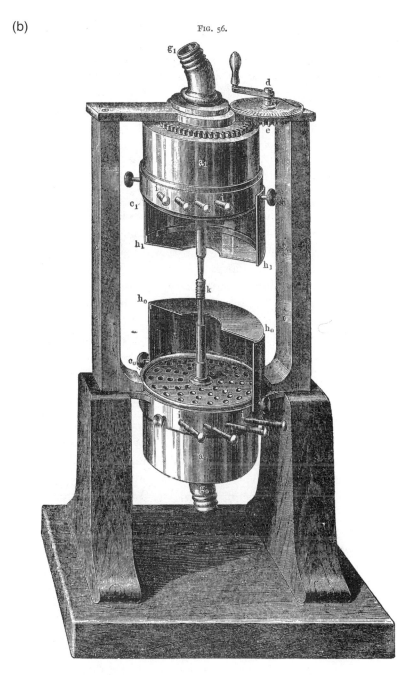

Figure 11.9 (continued)

In the years after Helmholtz's treatise appeared, Schoenberg and his circle devoted special attention to the issues of timbre that Helmholtz had studied and illuminated, as Julia Kursell has discussed.[7] Schoenberg coined the word *Klangfarbenmelodie* (literally "sound-color-melody") to describe the way timbre functions in musical composition, especially timbre's ever-growing role in orchestration beginning with Joseph Haydn.[8] Schoenberg's orchestral piece "Farben" ("Colors," the third of his Five Pieces for Orchestra, op. 16, 1909) is almost completely devoted to timbre. Instead of harmonic or melodic activity, a fairly constant harmony shifts slowly and subtly through various timbres, whose changing colors are the real substance of the composition. Schoenberg at times overlays several distinct layers of timbre, experimenting with this kind of timbral polyphony and the dreamy haze it conjures up (figure 11.10; ♪ sound example 11.7).[9]

Paradoxically, as they reached out past the late-Romantic sonic world, Schoenberg and others found themselves more interested in the polyphonic devices of Bach and even earlier composers. Anton Webern, one of Schoenberg's closest associates, was in fact a professional musicologist (then a newly scientific field of academic study), an expert in the Renaissance polyphony of Heinrich Isaac. To show his admiration of this work (and to promote awareness of it by a broader public), Webern even orchestrated Bach's *Ricercar à 6* in an innovative way that brought forward the new timbral possibilities of polyphony. To be sure, many other composers in the nineteenth century had orchestrated Bach, in various ways more or less assimilating his works to the orchestral sonic ideals of their time. But Webern took a much more radical approach: to each of Bach's phrases he assigned not just one but several different successive instruments. In so doing, he showed his interest in Schoenberg's concept of *Klangfarbenmelodie*: The same pitch played by a trombone, a flute, or a violin thus has a different "color," characteristic of each instrument.

For instance, consider how Webern refracted Bach's theme into no fewer than seven distinct orchestral timbres (figure 11.11; ♪ sound example 11.8), most of them just involving a single instrument (though generally within the brass family: muted trombone, horn, trumpet, with a touch of harp). In so doing, Webern did not proceed arbitrarily, merely applying orchestral color capriciously to Bach's lines; figure 11.11 shows how he assigned the different timbres within the theme precisely to reflect the divisions between its diatonic and chromatic aspects.[10]

In part, Webern was responding to the invitation implied by Bach's open score, which seemed to suggest that this monumental work might be conceived as much for an instrumental ensemble as for solo keyboard. Webern's timbral refraction of the *Ricercar à 6* also reflected his own instrumental style, his characteristically hypersensitive choice of very soft dynamics and continual dynamic shading: thus, he rendered the theme pianissimo, rising at most to piano. As Webern used such subtle nuances in the iridescent sounds of his own music, he presented Bach as also practicing a kind of "timbral polyphony": each voice in the ricercar is no longer a single entity but a changing sequence of several timbres.

Figure 11.10
The polyphony of timbres in Arnold Schoenberg's "Farben" ("Colors"), the third of his *Five Pieces for Orchestra*, op. 16 (1909) (♪ sound example 11.7).

Figure 11.11
Anton Webern's orchestration of the opening of Bach's *Ricercar à 6* (♪ sound example 11.8). A muted trombone plays the first (diatonic) element of the theme; the second (chromatic) element is passed sequentially between muted horn, trumpet, and trombone; the harp adds its timbre to accent the middle of this element; the third (closing) element uses horn, trumpet, and harp. In measure 6, the overlap between horn and trombone notes allows the fusion of their timbres.

Indeed, Webern's account of the historical development of polyphony invoked the concept of *dimension* by asserting that "when several parts sound at once, the result is a dimension of depth. … The idea is distributed in space, it isn't only in one part."[11] The claim that polyphony involves a new dimension of music (perhaps already implicit in Oresme's discussion of the dimensions of sound) is made explicit in Webern's presentation of "the *space* a musical idea can occupy" when several parts sound at once.[12] Though he did not discuss timbre as another dimension, Webern's timbral recomposition of Bach seemed to add new dimensions not just in number but in kind: attending to voices moving between timbres has a different felt quality than hearing voices each having a constant timbre.

Though such expanded use of timbres gave new coloristic dimensions to twentieth-century polyphony, in different ways composers continued to push the limits of the independence even of voices having a single timbre, for instance by using polytonality, the simultaneous use of two or more different tonal centers.[13] Still, ever since medieval polyphony, however independent the lines might be in language, style, or melody, they remained rhythmically correlated by a single underlying beat shared by all the voices. Even when Mozart included three simultaneous orchestras in the finale to act 1 of *Don Giovanni*, each playing a different dance—the minuet for the aristocrats, the contradance and waltz for the plebeians—all three moved to the same underlying pulse (figure 11.12; ♪ sound example 11.9). Tracking the development of more rhythmically diverse polyphony, György Ligeti noted that "Mahler was a keen observer of how the sound of various brass bands, roundabouts, and musical automata blended at a fair; he thought that was real polyphony. And it is what you find in the development section of [the first movement of] his Third Symphony."[14]

In the twentieth century, a number of composers set out to break this unanimity of pulse. Here (as in so many other ways) Charles Ives pioneered. In *The Unanswered Question* (1908), three instrumental groups play at fundamentally different tempos. A pianissimo ensemble of muted strings play continuously in a very slow 4/4 time, representing "the silences of the Druids—who know, hear, and speak nothing." Against the background of their solemn (and tonal) hymn, at certain moments a solo trumpet softly plays "the Question," a mysterious unresolved melody whose rhythm is constructed to suggest that it follows a different slow tempo, tonality, and meter. Later, a group of four wind instruments interject dissonant "Answers" in increasingly frenetic (and quite unrelated) tempos (figure 11.13; ♪ sound example 11.10a,b). By layering these various tempos, meters, and timbres, Ives sought to convey the fundamental incommensurability between the mysterious questions that surround human life and the busily assertive (though mostly empty) attempts to answer them advocated by conventional systems of belief. By comparison, the muted strings have a unified pulse, so that the ear gravitates to them as a force of continuity; even so, the trumpet's unanswered question haunts and adds mystery to the strings' serene hymn.

Other composers took the possibilities of polyphonic tempi and meters in different directions. Karlheinz Stockhausen's *Gruppen* (1957–58) used the spatial separation between three separate orchestras to underline the ways they might play in different tempi and metronomic rates under three different conductors, who intercoordinate only at certain sonic events that serve as common signposts or points of arrival. Unlike Ives, in this work Stockhausen had no explicit program or philosophical agenda. His listeners, surrounded by these three orchestras, experienced a new degree of polyphonic independence between the different ensembles, experiencing each orchestra playing to its own conductor's beat. Instead of tightly coordinated "lines" or harmonic progressions, the hearers perceived "moments," as Stockhausen and others came to call them: transient but arresting constellations of pitches and timbres that might serve as alternatives to the traditional drama of musical exposition, development, and reprise.[15] These moments were more or less isolated sound-events that were sometimes encounters between the different orchestras, sometimes happenings within a single one. Their punctuated quality emerged from the essential absence of a connecting beat, whose underlying rhythm (in contrast) had given such important grounding to so much previous polyphony.

Though his own style was very different, György Ligeti studied *Gruppen* closely as he shaped his own technique of "micropolyphony," in which "you hear a kind of impenetrable texture, something like a very densely woven cobweb" produced by a multitude of voices; though strict, this polyphony "remains hidden in a microscopic, under-water world, to us inaudible … you could say that there is a state of supersaturated polyphony, with all the

Figure 11.12
The end of the scene with three orchestras from the finale to act 1 of Wolfgang Amadéus Mozart's *Don Giovanni* (♪ sound example 11.9).

Figure 11.12 (continued)

Figure 11.13

Figure 11.13 (continued)

Charles Ives, *The Unanswered Question* (1908): (a) the beginning of the work, showing the strings as "the Druids" and the solo trumpet stating "the Question" (♪ sound example 11.10a); (b) the end, including "the fighting Answers" (♪ sound example 11.10b).

'crystal culture' in it but you cannot discern it. My aim was to arrest the process, to fix the supersaturated solution just at the moment before crystallization."[16] For example, Ligeti described his orchestral piece *Atmosphères* (1961) as "just a floating, fluctuating sound, although it is polyphonic."[17] His *Poème symphonique* (1962) for one hundred metronomes carried this polyphonic technique even further (figure 11.14; ♪ sound examples 11.11a,b). The metronomes are all wound up, set to different tempos, and then let run, beginning as simultaneously as possible before the audience enters the hall. At the beginning, the barrage of the metronomes' clicking sounds like a heavy rainstorm, with no single rhythm perceptible in the dense overlay of their sounds (♪ sound example 11.11a). At this level of density, all the different tempi merge into a single texture. After a while, the metronomes begin to run down, one by one, so that finally only a few remain and discernible patterns of interference between their beating emerge, which grow ever simpler and clearer until finally only one metronome remains clicking, the solitary victor, until it too runs down (♪ sound example 11.11b). Thus, at the opposite ends of tempo overlay there is a kind of simplification, the initial single texture versus the final single tempo.

Stockhausen depended on the coordinative beat of a conductor to give a unifying tempo to each separate orchestra, which thereby remained tied to old standards of instrumental cohesion and coordination; Ligeti's *Poème* plays with the mechanicality of its "voices."[18] Conlon Nancarrow used the player piano to create new kinds of polyrhythm, the mechanical instrument allowing degrees of control difficult for human performers. For instance, his Study no. 33 (composed approximately 1965–69) presented a series of canons at the ratio $\sqrt{2}/2$, the first effort in the history of music to use irrational tempo relationships; his Study no. 40—Canon e/π (1969–77)—used the ratio of two transcendental numbers, which are irrational and yet not the solution of any finite algebraic equation with integer coefficients.[19] As Oresme already realized, such irrational relationships as $\sqrt{2}/2$ would mean, strictly speaking, that such a canon, after its initial synchronization, would *never* again align with itself.

Because all mechanical devices have finite precision, no physical mechanism could ever realize such a canon precisely. In practice, the irrational relations must be rounded off; for his Study no. 33, Nancarrow approximated $\sqrt{2}/2$ within 99.97 percent as the ratio 41:29.[20] Even thus approximated, his canons convey the startling quality of this irrational relationship, in which nothing ever seems to "line up" after the initial chord, a quality he emphasized by contrasting sustained with staccato chords at the beginning of the piece (figure 11.15; ♪ sound example 11.12). The overall impression of "free," almost wildly uncoordinated playing comes from the irrational tempo relationship, giving a new depth of meaning to the concept of "irrationality." Though Nancarrow initially doubted that human players could ever render his scores accurately, in 2007 the Arditti Quartet recorded a string quartet version of his Study no. 33.[21] As difficult as that must have been, they accomplished this without using headphones and prerecorded click tracks to keep the

Figure 11.11
György Ligeti with some of the one hundred metronomes for his *Poème symphonique* (1962) (♪ sound examples 11.11a,b).

Figure 11.15
The opening of Conlon Nancarrow's Study no. 33—Canon $\sqrt{2}$ / 2 for player piano (1965–69) (♪ sound example 11.12).

players on their separate tempi, to which other ensembles resort when faced with such challenges.

For listeners as well as players, works such as Nancarrow's strain their ability to hold together voices whose relation is really incommensurable, struggling to hold them together, sometimes feeling the strange sensation of incoherence more akin to trying to listen to two spoken conversations at once. We seem at the limit of coherent polyphony; yet other contemporary composers were pushing that boundary even further, as we shall see in the next chapter.

12   Contrapuntal Radio and Polyphonic Fields

We can extend polyphony further to include "voices" that go far beyond the traditional limits of music, certainly beyond the limits of the ever-more-expressive musical tradition that suffered a kind of implosion through its own inherent instability. If we look toward the alternative legacy of music as science, the contrapuntal radio compositions of Glenn Gould explored the multiphony of several simultaneous speakers. Going further still, John Cage enlarged the domain and meaning of polyphony to include a continuum of "voices" that transformed the meaning of sounds and silence. The concept of "field" provides a helpful generalization of distinctly separate "voices" to a fluid continuum in which an indefinite number of elements freely mingle yet still preserve the fundamental tensions in the ways they coexist and interact.

After he stopped giving live concerts as a pianist, renowned for his intense and idiosyncratic interpretations of Bach, Gould constructed several extended works of what he called "contrapuntal radio."[1] His *Solitude Trilogy* tried a number of approaches. *The Idea of North* (1967, the first in his trilogy) begins with four spoken voices entering one by one, their statements overlaid "like a kind of trio sonata texture," as Gould described it (♪ sound example 12.1).[2] The various speakers are travelers on a train bound for the far North, which fascinated Gould as the permanent frontier that gives Canada its special character.[3] Each voice speaks conversationally, describing one or another aspect of their experiences of the North or telling stories; each is so engaging that when another voice starts speaking at the same time, one is unsure whether to continue following the first speaker or switch attention to the next. I at least could never understand both at once, much less when three or four spoke simultaneously, finding my attention flitting rapidly between them. To help the listener, Gould would bring up the level of one speaker compared to the rest (often the voice that has just entered), while bringing down the others. This is, in fact, the technique called "voicing" that he often used in his piano playing, especially for fugues: shaping the relative loudness of the polyphonic voices so as to draw attention to one of them more than others.

As if recognizing the difficulty of polyphonic speaking, Gould seldom overlays voices and then not for long. For long stretches, one hears only one voice at a time; during those passages, Gould seems to present most of the narrative content of his program: the difficulties faced by those who live in the far North, their various stories, satisfactions, dilemmas. Most of the speakers are decidedly "characters" in their own right, each of whom Gould allows their time in the sun, rather than imposing a fixed fugal form on them. When one voice cedes to another, Gould will sometimes fade between them, sometimes cut without fading. The moments of overlay between voices have an interesting effect, suggesting that no single voice (however intriguing) by itself can tell the full story of the North. By blending each voice with the others, all become parts of a larger unity, conveying a broader narrative than any of them would apart from the others.

After the initial "exposition" of the four voices, Gould makes a few introductory remarks and describes the speakers, but there is no narrator in the ordinary sense.[4] Though he describes a fifth character (Wally McClain, a particularly loquacious old-timer) as the program's narrator, this only means that McClain is heard more often than any of the others and has a central role among the characters. This avoidance of a controlling narrative voice avoids the "talking head" monotony of conventional documentaries and underlines the musical implication that we are experiencing a theater of voices. To create a larger sense of sweep, Gould often adds a background track, generally the sound of train cars traveling along the rails, sometimes accompanied by the clink of dishes in the train's restaurant car. He described this technique as a "continuo," alluding to the Baroque practice noted above of undergirding a composition with a continuous harmonic background.[5] As with conventional radio programs or films, this track adds ambience, here of the train the characters travel on their journeys north. Gould sometimes overlays two spoken voices against these ambient sounds.

As *The Idea of North* concludes, McClain expresses with considerable eloquence the sentiment that the struggle with the extreme North represents "a moral equivalent of war" against "Mother Earth," requiring a similar courage to that exhibited by soldiers but without bellicosity or violence harmful to humans. To underline the triumphal quality of this realization, under McClain's peroration Gould overlays the final movement of Sibelius's Fifth Symphony, which is both harmonically static (revolving around the same sequence of chords over and over, a kind of ostinato), yet gradually growing in volume and grandiosity (figure 12.1; ♪ sound example 12.2). The choice of Sibelius reflects his identity as a northern composer (though Finnish rather than Canadian), whose piano works Gould also recorded, but the passage in question functions more as a musical synthesis of the spoken voices (particularly McClain's), conveying their devotion to the struggle with the North.

As successful as the ending of *The Idea of North* may be, by resorting to time-tested formulas of sound track underlay, Gould seems to suspend the premise that spoken voices could form a satisfying contrapuntal whole by themselves. Still, as with all cinematic "sound tracks," his use of music reminds us that the polyphony of spoken voice against music is peculiarly effective, attested by its universal use in films and radio dramas. What

Figure 12.1
The ostinato from the final movement of Jean Sibelius's Fifth Symphony, op. 82 (5 after letter O, ♪ sound example 12.2), which accompanies the final section of Glenn Gould's *The Idea of North*.

the spoken voice cannot convey, music can provide. Such practices have a long history, reaching back to the "melodramas" of the early nineteenth century by Schubert and others, in which a piano accompanied a spoken text (figure 12.2; ♪ sound example 12.3).[6] Moreover, though several speakers may not cohere, speech and music can operate synergistically, not just separately; their artful overlay creates effects that neither could achieve separately. For instance, Gould times the final staccato chords of the Sibelius so that they come just after (and decidedly underline) the climactic spoken statement identifying "the journey North" with "the moral equivalent of war." Thus, the final musical cadence decisively closes the spoken program.

In 1970, the director Judith Pearlman made a film version of *The Idea of North* that may help clarify how contrapuntal radio affects the mind by comparison with the effect of visual images. Gould's original radio program was preserved as the soundtrack of the film, which shows him delivering the introduction (dressed appropriately in warm clothing, which he tended to wear even in temperate climates; figure 12.3a) as well as scenes with the various other voices, generally timed to coincide with their appearance in the radio version. The main addition the film makes is an unspeaking character, an unnamed young man (figure 12.3b) who is depicted making the journey north on the train, eventually falling into extended conversation with Wally. This character presumably grows out of Wally's own remarks in the radio program about a hypothetical young man going north, but seeing a particular dark-haired person in the film gives a kind of specificity to the spoken version.

Both that specificity and the cinematic imagery in general distract the viewer from the contrapuntal aspects of the radio program; the cinematic narration simplifies and clarifies the story to such an extent that one scarcely notices that the voices sometimes speak at the same time. After all, trying to imagine a visual analogue of polyphony is peculiarly difficult: however many visual elements might stand before the eye, they all form a (more or less) coherent composition that the mind "absorbs in one glance," even though the eye actually darts around the visual field (in "saccadic" motion), as extensive investigation has clarified. The suppression of this discontinuous saccadic activity seems important in relation to the apparent continuity of the final visual effect. For the moment, we should remember that film itself involves an essential discontinuity in time, that the projection of about twenty-four frames per second creates the felt appearance (or illusion) of continuous motion. These issues of continuity versus discontinuity are central, as we shall shortly consider.

In *The Latecomers* (1969), the second work in the *Solitude Trilogy*, Gould uses sound underlay continuously.[7] His protagonists live in Newfoundland, to which human habitation came late and remains a struggle against isolation and climate. To evoke life on this fairly remote island, Gould surrounds his speakers with the sound of the sea, often so loud that one struggles to understand their words against its ceaseless roar. As a result, the "continuo" of sea sounds seems to fight with the voices for the listeners' attention.

Figure 12.2
The beginning of Franz Schubert's melodrama "Abschied von der Erde," D. 829 (1826). The spoken text is placed above the piano part, indicating when in the musical context it is to be delivered (♪ sound example 12.3). Text: "Farewell, thou beautiful Earth! I can only now understand how joy and suffering come to us."

(a)

Figure 12.3
(a) A publicity photo by Glenn Gould for the film version of *The Idea of North* (1970); (b) the actor Geoffrey Reed as the "Young Man" in the film.

(b)

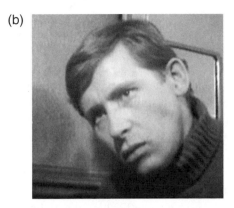

Figure 12.3 (continued)

Gould also moves the sea sounds back and forth between the stereo channels, further attracting the hearers' attention, sonically embodying the way the sea entirely surrounds the island.

In *The Latecomers*, Gould entirely omits any introduction of the characters or words of his own. The speakers are never named or otherwise identified than by their own voice and tone; even their number does not seem rigidly defined, though four of them are particularly prominent. Gould also seems to choose a different contrapuntal format: not a trio sonata (as in *The Idea of North*) but what one might call a two-part invention, in which generally one voice speaks, only occasionally overlaid against another voice at the same time, though always against the background continuo of the sea. To select the moments of transition of overlap between voices (sometimes in mid-sentence), Gould tends to choose passages in which one voice either continues the sentiment of another or pointedly contradicts it. Such moments of disagreement calls attention to the relative musical or timbral qualities of the voices. The only female speaker has a hauntingly monotonous voice, almost chanting her words on the A below middle C, seemingly unintentionally; so lulling is her speech that it is hard to maintain attention to the meaning of her words (♪ sound example 12.4). One of the male voices likewise is notably monotonous (curiously, on the same pitch, as if it were the tonality of the program). In contrast, another elderly speaker stands out for his colorful voice, which suits his marked character as a crusty yet eloquent advocate of the old ways, as does his charming Irish accent, colorful vocabulary, and wide pitch range.

The final work in Gould's trilogy, *The Quiet in the Land* (1977), concerns not geographical separation (as did the first two works) but more intangible issues of isolation embodied in the situation of the Mennonites, who traditionally aspired to be "in the world but not of it." Historically, Mennonites lived in separate communities, dressing and comporting themselves differently from those around them, not drinking, going to movies, or amusing themselves in "worldly" ways. Gould also evokes the divergent lives of those Mennonites who move away from traditional ways and those who try to remain faithful to them.

In this program, Gould's contrapuntal technique might be analogized to a three-part invention, for often three voices sound at once, generally two of sound or music, one of speech. The "continuo" here is a Mennonite service: the program begins with the bell that calls them to church, counterpointed against the sound of children playing and the occasional sounds of passing cars (as emblems of contemporary life). These are not artificially overlaid sounds but simply the manifold sounds of life, with its continual polyphony (to which we will return shortly). At intervals throughout the program, Gould returns to this service, as if it were continuing in the background, even if not clearly or consistently audible. Near the beginning, he includes an entire hymn sung by the congregation, counterpointed against various voices expressing the difficulties of withdrawing from or engaging with the world. As it progresses, a third voice enters, Janis Joplin singing "Oh Lord, Won't You Buy Me a Mercedes Benz?" alternating with a passage from Bach's fourth cello suite, another contrast between "secular" and "sacred" music, though all these musical works (including the hymn) share the common tonality of E-flat that seems the key of the whole program.

In *The Quiet in the Land*, Gould seems to reach a new level of comfort with the challenges of spoken counterpoint, probably as a result of his experiences with the prior two parts of the trilogy. He includes far longer and more frequent passages of two or three voices speaking simultaneously, passages that work partly as a result of the pauses each speaker includes, which allow the hearer's mind to travel back and forth between voices more comfortably. This is particularly true of the Mennonite service, whose pastor is particularly eloquent (in a simple, direct way) and who makes many thoughtful pauses, which Gould exploits in his polyphony. As Gould brings forward the voices of those wandering into worldly paths, the pastor calls the children to come forward and begins his sermon on the parable of the prodigal son. Though the sermon is not directed against modernizing Mennonites, the listener cannot help reflecting on their path in the context of the parable.

Just as these nonconformists turned away from their church's rigor, the program for a time leaves the sermon behind; the background music is now piano-bar jazz and even striptease touting. For the first time in his trilogy, Gould explicitly addresses his own art by including a Mennonite composer and other voices interested in the artistic possibilities opened by modernity, compared to the restrictions of their tradition. Gould includes also a striking passage in which a children's choir and harp are recording a composition, including the irritable comments of their conductor, reflecting Gould's own deep interest in the process of recording (♪ sound example 12.5). In this section, Gould uses particularly long stretches of two overlaid spoken voices to evoke the ambiguities raised by the conflict between hallowed traditions and modern freedom, a conflict that his contrapuntal radio reflects and projects with particular success. In contrast, he returns at the end to the church service and the pastor's sensitive homily to his congregation in preparation for communion. Whether consciously or not, Gould tends to use a single spoken voice to describe such wholehearted states, as against the divided, polyphonic consciousness more characteristic of the modern mind; one wonders whether he knew Søren Kierkegaard's *Purity of Heart Is to Will One Thing* and its critique of "double-mindedness."[8] Gould's own personal affection for the special calm of Sunday afternoons, the Anglican Church music of his youth, and "the peace that passeth all understanding" seems to underlie his sympathetic treatment of the Mennonites' service.[9] His program ends, as it had begun, with the tolling of a bell, now accompanied by the chatting of the congregants as they return to the world. Yet this bell sounds C#, dissonant against so much E-flat major; in light of Gould's careful observance of tonality, this dissonance may be emblematic of fissures within the Mennonites' ideal world.

Listening to Gould's contrapuntal radio programs, the ear seems to move its attention from voice to voice with discontinuous jumps, as the eye does in scanning a visual image. Because they present a series of extended examples that challenge the listener, the experience of hearing his programs gives further evidence for the basic intuition noted in the case of listening to simultaneous conversations, that one's attention shifts back and forth between them more or less rapidly. Still, the eye actually performs its saccadic motions in various spatial directions, whereas the ear does not, so that the discontinuity of jumping between voices as well as the apparent continuity of musical polyphony must come from processing upstream from the auditory nerve in some higher centers in the brain. Before considering what may be happening at the level of brain processing, though, we will need to return to the felt qualities of continuity that characterize our experience of many sounds, as they do of our visual experience.

Probably the most extreme—and hence most thought-provoking—extension of polyphony came in the work of John Cage, who disclaimed "expression" and opened the simultaneous hearing of many seemingly unrelated sounds in ways unprecedented in the preceding history of music.[10] Though he "had been taught, as most people are, that music is in effect the expression of an individual's ego—'self-expression,'" in the mid-1940s Cage decided that "everyone was expressing himself differently" so that "no one was understanding anybody else; for instance, I wrote a sad piece and people hearing it laughed."[11] Cage found this "pointless" and stopped writing music until he found a new approach he drew from Eastern traditions but also from the seventeenth century English composer Thomas Mace: "The purpose of music is to sober and quiet the mind, thus making us susceptible to divine influences."[12] Cage thus reengaged with the dispassionate, ecstatic musical tradition we considered in Evagrius of Pontus.

Over the years, Cage came to think of these "divine influences" as "all the things that happen in creation," for "many things, wherever one is, whatever one's doing, happen at once. They are in the air; they belong to all of us. Life is abundant. People are polyattentive."[13] His "Where Are We Going? and What Are We Doing?" (1961) presents texts to be delivered as four simultaneous lectures; for instance, one lecture may be spoken live while the other three are played from prerecorded tapes.[14] Well before Gould tried contrapuntal radio, Cage took a much more radical approach, composing his texts using his *Cartridge Music* (1960), itself based on indeterminate graphic notation.[15] As a result, the number of voices to be heard simultaneously varies without any perceptible order; there is no attempt to "voice" them (as Gould had) or stage-manage their overlay dramatically. Indeed, in one extended passage of "Where Are We Going?" all four voices happen to fall silent. Each text is perfectly intelligible on its own, as the listener notices in those passages when only one is speaking. But often the overlay makes them no longer understandable in any ordinary sense; the listener jumps back and forth between voices, occasionally grasping some words or phrases from one or another.

Accordingly, Cage even contemplated printing all four texts completely superimposed over each other but rejected that as "a project unattractive in the present instance."[16] Though separating them graphically made them "legible," he considered that "a dubious advantage, for I had wanted to say that our experiences, gotten as they are all at once, pass beyond our understanding." Following a traditional Indian concept that the "function of the artist is to imitate nature in her manner of operation," Cage wished to put nature "in the driver's seat," rather than relying on human control, so that this lecture would help us "see meaningless-ness as ultimate meaning" because, in its overlaid voices, "meaning is not easy to come by."[17] Cage concludes by inviting his listeners to pass beyond "understanding": "Here we are. Let us say Yes to our presence together in Chaos."

During the same period, Cage used the medium of radio to create even more indeter-minate polyphonies.[18] For instance, each performer (anywhere from one to eight) in his *Radio Music* (1956) operates a single radio by playing one of fifty-six different frequen-cies on the AM band (including "——" for "silence") according to a list Cage provided for each part (figure 12.4), there being no overall score (as with the Renaissance polyphonic works that consisted of separate written parts). Each performer is supposed to make a list of durations (ideally prepared using chance operations) during which they will play those stated frequencies on their radio so that the whole piece will last six minutes total. Assuming the maximum number of performers (eight), the piece will be whatever hap-pens to be on the air at that time, according to the list each performer follows. The same quality would characterize a solo performance made using one radio following its own score (presumably randomly chosen from the eight parts Cage provided, such as ♪ sound example 12.6).

The polyphony that results is far more radical in character than any we have considered so far because no single mind (neither the composer nor the performer and certainly not the listener) can know in advance what will happen, apart from the radio frequencies to be se-lected. Even a "dress rehearsal" before the performance would produce an entirely differ-ent montage of sounds. No tempo or plan (other than the stated frequencies and durations) coordinates the result. Indeed, simultaneous performances of this work in different loca-tions (with different radio stations at the given frequencies) could sound utterly different.

PART A OF RADIO MUSIC to be played alone or in combination
with Parts B-H. In 4 sections (I-IV) to be programmed by
the player with or without silence between sections, the
4 to take place within a total time-length of 6 minutes.
Duration of individual tunings free. Each tuning to be
expressed by maximum amplitude. A _____ indicates 'silence'
obtained by reducing amplitude approximately to zero. Before
beginning to play, turn radio on with amplitude near zero.

JOHN CAGE
STONY POINT, N.Y.
MAY 1956

I	(I cont.)	(IV cont.)
105	107	91
———	———	———
125	69	146
55	107	69
	II 56	———
———	56	
91	124	97
60	125	——
69	———	91
76	120	156
112	55	——
56	56	55
———	125	155
86	69	128
73	84	——
127	**III** 120	138
73	———	——
148	76	107
76	———	——
109	**IV** 99	99
63	———	——
67	69	153
91	———	63
86		
73		

Figure 12.4
One of the eight possible parts that John Cage provided for *Radio Music* (1956) (♪ sound example 12.6).

Listening to several performances of this work, one can be treated to a whimsical montage of sports, pop songs, talk shows, and (rather often) just static. Reflecting the nostalgic aura of radio (now absorbed by podcasts), even the possibility of performing *Radio Music* seems on the verge of disappearing. Digital radio has replaced analog in many countries; as this happens, Cage's radio works will then no longer be performable because digital radio does not allow the selection of specific frequencies, as he specifies, only entire channels. Appropriately for this most American of composers, his native airwaves remain open to his anachronistic radios, at least for the time being.

The curious status of "silence" in these works deserves attention. Even though Cage specifies that it means "reducing amplitude approximately to zero," what results is not exactly a "rest," in the traditional sense that began with the rhythmic notation of the Middle Ages devised by Oresme's friend Philippe de Vitry. During such a "rest," the singer does not sing; during Cage's "silences," the radio continues to operate, still playing (albeit very softly) whatever is on the air at the time. Cage went on to give much attention to "silence," even writing a famous "silent" piece, *4′33″* (1952), which he considered his most important work. It comprises three movements, each marked "tacet," the standard musical term directing that an instrument or voice does not sound. Here is the entirety of Cage's specification (in lieu of a "score"): "The title of this work is the total length in minutes and seconds of its performance. At Woodstock, N.Y., August 29, 1952, the title was 4′33″ and the three parts were 33″, 2′40″, and 1′20″. It was performed by David Tudor, pianist, who indicated the beginnings of parts by closing, the endings by opening, the keyboard lid. However, the work may be performed by any instrumentalist or combination of instrumentalists and last any length of time."

Of that first performance, Cage recalled that "you could hear the wind stirring outside during the first movement. During the second, raindrops began pattering the roof, and during the third the people themselves made all kinds of interesting sounds as they talked or walked out."[19] What the audience heard during the duration of the work was the ambient sounds of their environment—birdsong, street sounds, rainfall, coughs—presented to them as a composition, rather than those sounds being the more or less ignored background for a "real" piece of music played on that piano.[20] For Cage, this was a polyphonic experience because "by silence, I mean the multiplicity of activity that constantly surrounds us. We call it 'silence' because it is free of *our* activity. It does not correspond to ideas of order or expressive feeling—they lead to order and expression, but when they do, it 'deafens' us to the sounds themselves."[21]

Cage later explained that his inspiration was the "white paintings" of Robert Rauschenberg, in which what the viewer sees are the shadows or reflections that happened to fall on the paintings' blank white surfaces. Also, Cage often referred to his visit to the Harvard Anechoic Chamber, a space deadened by thick absorbent padding, designed in 1943 to investigate sound in combat vehicles (figure 12.5). "In that silent room, I heard two sounds, one high and one low," which an engineer there interpreted for him: "The high one was

Figure 12.5
Harvard Anechoic Chamber, 1948 (built 1943, destroyed 1971).

your nervous system in operation. The low one was your blood in circulation."[22] Entering that chamber as a student in the late 1960s (a few years after meeting Cage), I too remember that the "silence" there seemed very loud, including high whistling tones and a whole spectrum of sound whose like I had only experienced during stays in the remote California desert.

Cage's interventions call into question much about polyphony (and indeed about sound and music in general). The number of performers or musical "voices" is no longer determinate: any one of the radios in *Radio Music* contains many "voices" and the listener to the "silent" piece hears many things with no clear sense (or even the possibility of knowing) from whence they come or whether they are really distinct. Because of that, it is no longer clear in what sense the concept of polyphony still describes these compositions or if indeed they have reached some maximal condition, a kind of continuum of voices. Consider further the experience of the "silent" piece: I might notice differences between the various sounds I hear, but at some point I may feel all of them as part of a single "composition" of which I am the sole auditor, everyone else present having their own unique experience, from their point of hearing and depending on what they hear (or to what they give their attention).

Nor does this demonstrate that my listening self constitutes some nuclear core of identity or what Immanuel Kant called an "original synthetic unity of apperception."[23] Perhaps this listening "I" might dissolve, as ego, into a pure receptivity in which the sounds hang suspended; or perhaps those sounds may bring into being that very receptivity (that "self," if you will), or at least the "self-that-hears-these-sounds." Here "listening" seems very different from mere passive "hearing": someone seated next to me who was more acute and perceptive might notice sounds I ignored, nor would there be a "score" against which we could later compare our experiences to determine "what really happened." Even if we recorded the occasion and played it back, already other sounds would have entered in than those we originally heard.

To clarify these implications, consider Cage's *Variations IV* (1963), designated "for any number of players, any sounds or combinations of sounds produced by any means, with or without other activities."[24] This work forms part of a trilogy, of which *Atlas Eclipticalis* (1961–2) was the first composition, associated with nirvana in Hidekazu Yoshida's interpretations of Japanese haiku poetry.[25] Second in the trilogy, *Variations IV* is associated with samsara, the turmoil of everyday life, followed by *0′00″* (1962), representing individual action. As score for *Variations IV*, Cage provided a transparency with two small circles and seven dots, to be cut up and dropped on a map of the performance space, thereby indicating the locations of the actions/sounds to be produced, though noting that "there are no indications of durations, dynamics, etc." (figure 12.6).

This radical openness to the kinds of action or sound continues the direction of *4′33″* but now adds a spatial element, modulated by the "random" placement of the circles and dots. As with much of Cage's other work based on the tossing of coins, *I Ching* divination,

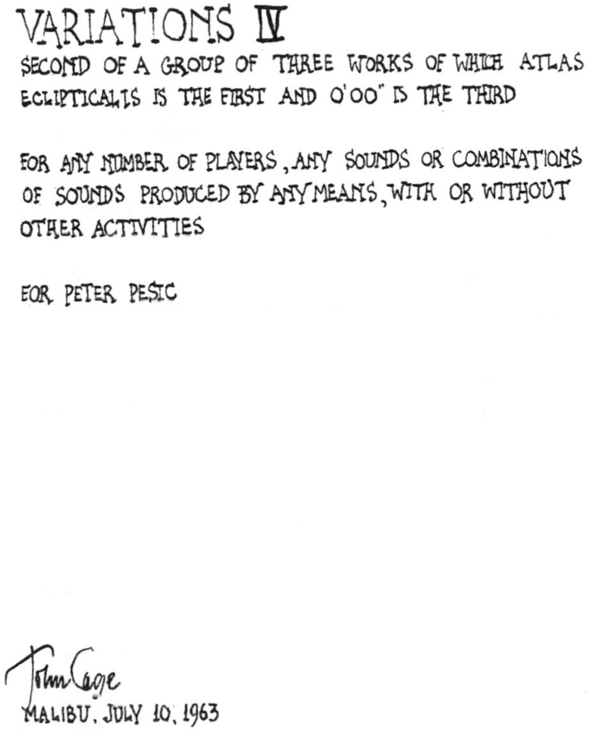

VARIATIONS IV
SECOND OF A GROUP OF THREE WORKS OF WHICH ATLAS
ECLIPTICALIS IS THE FIRST AND O'OO" IS THE THIRD

FOR ANY NUMBER OF PLAYERS, ANY SOUNDS OR COMBINATIONS
OF SOUNDS PRODUCED BY ANY MEANS, WITH OR WITHOUT
OTHER ACTIVITIES

FOR PETER PESIC

John Cage
MALIBU, JULY 10, 1963

Figure 12.6
John Cage, *Variations IV* (1963).

or star charts (to give only a few examples), the meaning of "randomness" is also opened for our consideration. The sounds we hear during *4'33"* are presumably random, but the way we take them into consideration, *listen* to them, may grant them a significance or quality that goes far beyond the dismissiveness implied when speaking of "merely random" events.

Though he expressly distanced himself from improvisation, as ordinarily understood, in *Variations IV* Cage asked his performers—and thereby also his listeners, for they cannot be absolutely distinguished—to listen to or create a multiplicity of sounds and events in a multiplicity of places.[26] In this work, the specification of *places*—determined by chance operations—precedes the choice of what sounds or actions to perform at those places. As Helmholtz observed, the ear is far more able to discern directionality of sound than the eye can do for sight.[27] By shaping the piece in terms of the spatial locations of its sound events (however transitory they may be), Cage seems to reconfigure polyphony from denumerate voices with no indication of their relative locations into a virtual sound-space. In that sense, the experience of *Variations IV* may be called *listening to a space* or *to a virtual spatiality.* When one listens to recordings of Cage and Tudor performing this piece in 1963, one is transported back in time to the Los Angeles gallery and the particular soundscape they evoked from that time and place.

Even more, Cage here includes dance as explicitly part of music, rather than one art subordinated to the other. Not only does his score put action on a level with sound, the first performance of *Variations IV* (at UCLA in July 1963) occurred as part of Merce Cunningham's *Field Dances*. For this work, Cunningham recalled that he was inspired by children in Boulder, Colorado, playing outside while he taught: "They were having such a beautiful time. *Field Dances* came from that because I could see that they were running and skipping and to me it was dancing; but for them it wasn't different."[28] Cunningham's title for his dance suggests both the pleasure of playing in an outdoor field but also may be read in terms of the larger connotations of *field. Variations IV* was not simply the music "accompanying" the dance; Cage understood that the actions of the dancers (and the sounds they made) occupied the same zone as the sound production that he and Tudor performed, along with the ambient sounds acknowledged in *4'33".* Further, as I recall from seeing this performance (and the surprise of learning that *Variations IV* was dedicated to me), there was no predetermined synchrony between musical and dance actions; this decision to avoid any overt or explicit coincidence between them reflected long-standing artistic policy in the collaboration of Cunningham and Cage.[29] Even so, music *was* dance, as Cage already indicated when he had Cunningham conduct the world premiere of his *Concert for Piano and Orchestra* (1957–58) as the grand finale of his twenty-fifth anniversary retrospective concert in New York (1958).

This deep relation between music and dance may inform our inquiry into polyphony, if we begin to read music and dance as actions in a field. Here, we should recall the definition of *field* in physics as a continuous zone crossed by invisible (but powerful) forces. Cage grew up with the concept of field: his father was an inventor who thought the universe was governed by "electrostatic field theory."[30] Cage considered his own treatment of musical scores as pictures to be a "by-product" of his intention "to make a notation that would recognize that sounds did truly exist in a field," for which "we needed other notation in order to let sound be at any pitch, rather than at prescribed pitches."[31] Further, the actions of a single dancer are already polymodal in the sense that they coordinate many parts of the body and move in a larger field, a zone that includes that body but goes far beyond it. Thus, one of Isadora Duncan's most powerful dances simply involved her rising very slowly from lying prone to standing erect; I remember seeing Rudolf Nureyev electrify an audience just by gazing at them and raising his left arm very slowly.[32] When several dancers are performing even seemingly unrelated actions (as in Cunningham's choreography), they still form part of a single field, however much they act independently.

That same concept of field may also describe the virtual zone in which polyphony unfolds, even when the "voices" are as independent as Cage's radios or the indeterminate number of performers of *Variations IV*. When the individuals are strongly correlated, the field has one character (as with monophony or chant), but even when they seem utterly divergent, they still form a field, though of very different character, having varying lines of relative tension or relaxation, correlation or disjunction. Thus, even a heterogeneous assemblage of voices or dancers sharing a single space find themselves in a field, even without a single consciousness that grasps them as a unity. The mere awareness of that field—the lack of any sharp divisions between "voices," even spatially—also marks a significant step beyond the discrete polyphony of well-separated and distinct sources. We will now consider to what extent that field may be associated with its own perception and perceiver.

IV

POLYPHONIC BRAINS

13 Polyphonic Selves

Growing out of the rich development of polyphonic music, many other fields drew on the concepts of polyphonicity and dissonance. For them, polyphony in the general sense of "many-voicedness" gave helpful language for describing a spectrum of simultaneous yet interconnected phenomena that became important in a variety of contexts. Reflecting on this wide range, this chapter will include examples from psychology, literature, and sociology involving multiple simultaneous processes under the rubric of polyphony. In nineteenth-century psychology, the "voices" within the human psyche could become so dissonant as to verge on being a collection of disunified selves. Such concerns grew with the development of neurology and growing awareness of the specialized roles of the cerebral hemispheres, increasingly connected with clinical cases of "multiple personalities." These investigations also often involved musical evidence. The development of "split brain" techniques allowed new access to the separate functioning of the cerebral hemispheres. In literature, Marcel Proust presented the inner drama of the many selves that live within each person. Amy Lowell and Mikhail Bakhtin each enlisted polyphony to express their new visions of poetry and narrative; Leon Festinger coined "cognitive dissonance" to clarify the felt inner conflicts that shape social psychology. In sociology, Max Weber presented polyphony as a central factor in the "rationalization" of Western music and society; Pierre Bourdieu and Bernard Lahire disagreed about the degree of coherence among the multiple aspects of each person's identity and social activity.

Beginning in the eighteenth century, the question of multiple persons or personalities within a single individual had an increasingly vivid and highly contested history. Indeed, Alfred North Whitehead judged that in the nineteenth century each individual was "divided against himself."[1] Henri Ellenberger concluded that this century "was preoccupied with the problem of the coexistence of ... two minds [within one person] and of their relationship to each other."[2] This problem has two distinct dimensions. The first extends Descartes's question about the dual hemispheres of the brain (and the duality of most sense organs) in relation to the seeming unity of sense perception. The second concerns the possibility of multiple personalities within a single individual.

In the wake of Descartes's controversial suggestion that the pineal gland, deep inside the brain, was the seat of the *sensorium commune* ("common sense") where all sensory input merged, eighteenth-century neurologists entertained a number of alternative possible organs for this role, but none had any immediate connection with consciousness.[3] Franz Joseph Gall was the leader of a movement later called phrenology that mapped the functions of the brain onto its parts, each of which served as a specific faculty for specific tasks and whose relative strength could be measured through examination of the cranium (figure 13.1).[4] Thus, phrenology aspired to be something far more than just a catalog of bumps on the skull, reflecting its underlying hypothesis that the different brain areas could be gauged by their relative size and hence effect on the surrounding bone.

Curious or even bizarre as it may seem now, phrenology was a serious attempt to localize mental functions in different regions of the brain. In the process, it unseated the Cartesian picture of a unitary soul that sat, like an immaterial homunculus, "expertly manipulating bits of cortex like a pianist seated at a keyboard," as Anne Harrington puts it. Instead, the way the mind dwelled in the brain resembled (as the German philosopher F. A. Lange put it in 1881) "a parliament of little men together, of whom, as also happens of real parliaments, each possesses only one single idea which he is ceaselessly trying to assert. ... Instead of *one* soul, phrenology gives us nearly forty."[5]

If so, the two independent hemispheres of the brain might function independently, even highly dissonantly. In *Man a Machine* (*L'homme machine*, 1747), Julien Offray de La Mettrie invoked conflicting cerebral lobes to describe how Blaise Pascal

always required a rampart of chairs or else someone close to him at the left, to prevent his seeing horrible abysses into which (in spite of his understanding these illusions) he sometimes feared that he might fall. ... Great man on one side of his nature, on the other he was half mad. Madness and wisdom each had its compartment, or its lobe, the two separated by a fissure.[6]

Gall recorded some patients who felt they were going mad on one side of their brain; the pioneering American psychiatrist Benjamin Rush described several cases in which patients acted as if they had "*two* minds," which he thought might be explained by Gall's idea that the mind was "a double organ, occupying the two opposite hemispheres of the brain."[7]

Over the course of the nineteenth century, further cases accumulated of persons torn by contradictory impulses in whom there seemed "two minds, one tending to correct ... the aberrations of the other," as Henry Holland put it in 1840.[8] Worse still, in a few other cases one cerebral hemisphere was found to be entirely absent, leaving a gaping hole inside the skull (as autopsy revealed), and yet the patient "had conversed rationally and even written verses within a few days of his death." Having witnessed such an autopsy, Arthur Wigan argued in 1844 that if one hemisphere could sustain a whole mind, two hemispheres implied a "duality of the mind."[9] Others hypothesized that the phenomenon of *déja vu* (or paramnesia), the eerie sense of already having lived through an experience, might occur

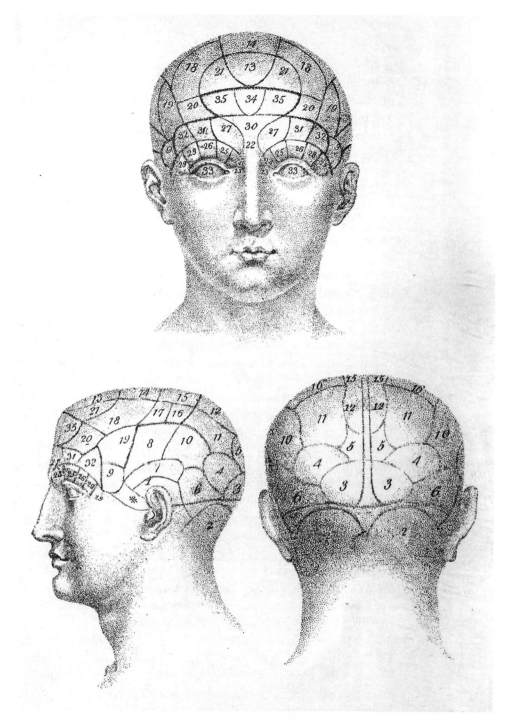

Figure 13.1
The phrenological model of the brain, from J. G. Spurzheim, *Phrenology or the Doctrine of Mental Phenomena* (1833). Note the corresponding numbering of corresponding faculties in the symmetrical regions of the two hemispheres.

as the two hemispheres "project double perceptions outwards beside each other in time."[10] Such hypotheses raised new questions about the multiplicity and localization of cerebral function, which now had to cope with a far more complex and dynamic picture of the brain than the static faculties of Gall and his followers.

Even so, Gall's basic vision of a map of localized brain functions (though not his palpable cranial bumps) stands behind Paul Broca's 1861 discovery of an area (now named after him) in the cerebral cortex responsible for "the faculty of coordinating the movements appropriate for articulate language."[11] Investigating further cases of aphasia, Broca noted that this area was located in the left hemisphere, rather than the right, thus violating long-standing expectations that the two hemispheres were entirely symmetrical.[12] Broca's work began the long process of clarifying how each hemisphere often controlled functions on the opposite side of the body: right-handedness comes from a dominant left hemisphere. Thus began the study of the relative dominance between the hemispheres, which then reopened the Cartesian question of what could unite the doubled brain.

Turning from neurology to psychology, by 1880 double personality became one of the most discussed mental disorders, though at the beginning of the century it had been considered very rare, if not completely mythical. The case of Mary Reynolds became a central example; about 1811, a taciturn and somewhat depressive young woman of nineteen, she fell into a deep sleep from which she awoke without any memory of her previous life, though she quickly regained lost skills. Her new personality was outgoing, almost reckless; even her handwriting was different. Weeks later she reverted to her initial state, now unable to remember the second. She alternated between these two distinct personalities until age thirty-five, after which she remained in the second.[13] Trying to offer scientific alternatives for spiritualistic or occult interpretations, contemporary physicians often chose to explain such cases in terms of the two hemispheres acquiring a pathological independence that led to double personality.[14] They also read striking asymmetries in the faces of the insane in terms of an underlying cerebral asymmetry, especially of the right hemisphere then considered the "mad" or irrational side of the brain (figure 13.2), a side that often was gendered as feminine. More broadly, the innate duality within a single mind was a deep preoccupation of literature, from J. W. von Goethe's Faust (1808), who confessed that "in me there are two souls, alas, and their / division tears my life in two," to Robert Louis Stevenson's Dr. Jekyll (1886), who learned "the thorough and primitive duality of man" from "the two natures that contended in the field of my consciousness."[15]

The double brain was increasingly also interpreted as enabling complex skills typical of high civilization, of which piano-playing became a favorite example. In 1879, the neurologist Jules Bernard Luys emphasized the polyphonic aspects of pianism: because each hand independently plays two different melodies, therefore "each [cerebral] lobe is capable … of acting separately and thus of generating a series of voluntary and conscious movements, inspired by a series of psychical operations, equally distinct on the right and left."[16] This separation of hemispheres does "not apply solely to psychomotor phenomena but also to

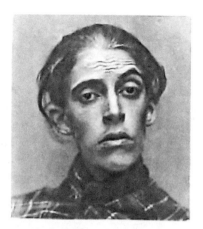

Figure 13.2
Two of John Turner's 1892 photographs of "asymmetrical faces of the insane." Note especially the left-hand side of the face, controlled by the right hemisphere, considered at the time to be the "mad" or irrational side of the brain.

mental operations, in order to read the music, gather memories, accomplish operations of judgment and coordinate motor activities."[17] Returning to the same question in 1888, Luys phrased his paradoxical conclusion even more sharply:

One arrives at this strange conclusion: when a pianist plays, mental unity is arrived at by splitting into two independent parts, appearing as if the left and right sides [of the brain] were so well separated that it seemed that the pianist had two separate sub-individualities that deliberate and act in isolation, as two instrumentalists play their parts in isolation.[18]

Even more, if the pianist also sings while playing, "one will really be struck with marvel and surprise at the infinite resources presented by that admirable instrument, the human brain," especially as offered by "the riches of harmony of the masters of our era" compared to the lesser counterpoint of "the ancient world and the Middle Ages." Here, the brain becomes an instrument no less marvelous than the piano it can play.

Though Luys emphasized the equal, independent play of separate hemispheres, his German contemporary Carl Wernicke argued that because "the novice piano player feels compelled to move the fingers of the left hand in synchrony with the right," therefore piano playing depended on adequate "innervation from one hemisphere alone," namely the left hemisphere that controlled the right hand.[19] Wernicke thus interpreted polyphony as a masked form of homophony, in which the right hand (usually playing the melody) really commanded the left hand merely as its accompanist. To be sure, home music-making in the nineteenth century often used just such a simple texture, left-hand chords supporting a right-hand melody; perhaps Wernicke was reflecting such common experience (he himself had taken up piano playing late in life).[20] Luys, in contrast, seemed to have in mind the independent voices of a Bach two-part invention, in which neither hand is simply predominant but each freely imitates the other. On both readings, polyphonic pianism required either overall control by one hemisphere or else a more mysterious coordination between two independent hemispheres.

Such concerns gave rise to novel educational movements, including "ambidextral training" for children to cultivate both hemispheres, not just the dominant one (generally the left). By 1900, two thousand Philadelphia schoolchildren were regularly engaged in such training, claimed to make them "relatively sharper and more intelligent" than their untrained fellows.[21] The training involved both hands—and hemispheres—in coordinated tasks, such as the ambidextral drawing illustrated in figure 13.3. As Harrington notes, this movement promised "a brave new world of two-handed, two-brained citizens, with untold benefits for health, intelligence, handcrafts, sport, schoolwork, industry, and the military."[22] Here again the arguments encompassed the polyphonic requirements of pianism; John Jackson, a champion of ambidextral training, envisioned that, "if required, one hand

Figure 13.3
A student in Philadelphia practicing "ambidextrous coordination in four directions," from J. Liberty Tadd, *New Methods in Education* (1899).

shall be writing an original letter, and the other shall be playing the piano … with no diminution in the power of concentration."[23] Even the critics who claim that "right-handedness is woven in the brain," Jackson notes, "commit their own children to the tender mercies of pianoforte teachers" precisely to cultivate "THE EQUAL SIMULTANEOUS DEVELOPMENT OF BOTH HANDS."[24]

In the eyes of the proponents of ambidextral training, piano-playing went far beyond merely cultivating a duality of hands or hemispheres. Because a Mendelssohn presto requires the pianist to execute over "200 intelligent volitions or movements per second, … even a dual or a triple mind would seem quite inadequate to accomplish so marvelous a task," in Jackson's judgment.[25] Because the brain "can maintain such phenomenal activity for one, nay for two hours at a time without apparent injury," Jackson concluded that practicing ambidextrous writing cannot be mentally dangerous, despite the fears expressed by its critics. On the contrary, he cites a testimonial by the celebrated juggler and music-hall performer Paul Cinquevalli, who recounted that he learned "to sit at a piano and play an accompaniment to my own whistling, composed of various tunes which were dictated to me by a person standing on my left; at the same time another standing on my right dictated a letter which I wrote down with my right hand."[26] Along with these simultaneous musical and digital skills, Cinquevalli claimed to have attained a superlative level of polyphonic awareness: "Also I could follow a conversation between two people, juggle two or three objects with my right hand, and follow a third person trying to puzzle me by rushing from one tune to another."[27] Thinking back to the problematic example of understanding multiple conversations, one wonders whether Cinquevalli could have followed a conversation between more than two people, whether or not accompanied by juggling or other activities; even he made no such claim.

Such attempts at ambidextral training rested on simple notions of cerebral symmetry that left unaddressed questions about the exact function of the nondominant (generally right) hemisphere. Clinical evidence eventually went beyond its earlier characterization as merely "mad" or irrational. To examine only one figure in this long development, in the last half of the nineteenth century the neurologist John Hughlings Jackson gradually came to the view that the left hemisphere served voluntary movements (including speech), the right voluntary sensation (including visual perception).[28] His early interests in music and philosophy seemed to have led to his particular turn to clinical medicine. As a young practitioner in 1866, he described cases in which the patient could sing but not speak, indicating that, though the muscles were intact, the brain areas involved were somehow only empowered for music rather than propositional utterances.[29] As such, these cases led beyond the work of Broca and Wernicke that had established the possibility that, though the production of audible speech may have been impaired by brain lesions, the underlying thought might yet remain intact in the patient's mind.

To interpret his cases, Hughlings Jackson turned to descriptions of music by the philosopher Herbert Spencer to distinguish between words and tones in speech, respectively providing "the signs of ideas and the signs of feelings. While certain articulations express the thought, certain vocal sounds express the more or less of pain or pleasure which the thought gives. Using the word *cadence* in an unusually extended sense, as comprehending all modifications of voice, we may say that *cadence is the commentary of the emotions on the propositions of the intellect.*"[30] Over time, Hughlings Jackson further refined this view to clarify that that such patients, damaged in the left but not the right hemisphere, were capable of singing the cadence of their feelings, though not of uttering a verbal proposition about them. No longer merely mute or irrational, the right hemisphere was thus indicated as the seat of musical and emotional expressivity.

In the twentieth century, a sustained reevaluation of the roles of the brain hemispheres led to several decades (about 1920–50) during which the whole topic of cerebral localization was decidedly suspect. Psychiatry, as an increasingly distinct practice, took a more complex stance toward neurology. On one hand, there remained the desire to ground psychology strictly in neurology, as with the young Sigmund Freud's *Project for a Scientific Psychology* (about 1895).[31] Yet the limited success of that approach at the time led psychiatrists to interpret mental conditions in more general ways than the earlier search for cerebral lesions localizing conditions in the brain. The "talking cure" and more holistic ways of understanding the mind (such as Gestalt psychology) seemed more promising, compared to the lack of any therapeutic help provided by purely neurological diagnosis: why try to pinpoint the afflicted part of the brain if that in itself did not provide any relief for the sufferer?[32] Indeed, the rich (and tragic) harvest of clinical evidence from the mental and physical traumas inflicted during World War I both aided and complicated the search to localize symptoms in the brain.

Freud's later view of the mind in some ways reflected the heritage of Plato's three-part soul, in which reason assigned itself authority over erotic and assertive desires, using music to soothe and guide them. Yet Freud's ego regulates the id much more insecurely; unconscious desires keep emerging even through the censorship of the superego, manifest in slips of the tongue and dreams, if not in outright action.[33] Describing Michelangelo's sculpture of Moses, Freud confessed that "with music, I am incapable of obtaining any pleasure. Some rationalistic, or perhaps analytic, turn of mind in me rebels against being moved by a thing without knowing why I am thus affected and what it is that affects me."[34] Dissonance seems the very texture of the mind, with ego and id inevitably at odds. Writing in 1914, the disillusioned Freud does not mention the Platonic vision of music as a science capable of regulating the passions, only his own incapacity; for Freud, what remains is to hold on to reason as Moses tried to hold on to the Tables of the Law even after witnessing the debacle of the golden calf.

In the years following the war, as its centers of research moved more and more to the English-speaking world, psychology increasingly sought to make itself into a fully objective and experimental science that focused on animals rather than humans, with their imponderable subjectivity. This turn to behaviorism emphasized observable actions, not unobservable inner states. Thus, Karl S. Lashley argued in 1926 that the loss of memory in rats depended only on the net mass of brain tissue removed, not its specific location. By 1929, he concluded that "specialization of function of different parts of the cortex occurs in all forms, but at best this is only a gross affair. ... The more complicated and difficult the activity, the less the evidence for its limitation to any single part of the nervous system ... [so that] there is little evidence of finer cortical differentiation in man than in the rat."[35] As Harrington put it, Lashley became "the intellectual leader of a movement in experimental neurophysiology away from a localizationist approach to cerebral functioning and toward a field theory of brain activity, a theory neatly encapsulated in two key terms: *equipotentiality* [of all regions of the brain] and *mass action* [meaning that cerebral power is only a function of the gross volume of the brain involved, not the particular region in play]."[36]

Nevertheless, experimental developments in the 1950s revitalized the localized theory of brain function. Wilder Penfield observed that electrically stimulating various points on the exposed cortex of the brain during surgery of patients who remained conscious would evoke specific memories, seemingly stored at those exact locations.[37] Indeed, Penfield himself described the brain as a "whole system [that] vibrates, one might say, with an energy that is normally held in disciplined control, like that of a vast symphony orchestra."[38]

Roger Sperry, who had been a graduate student of Lashley's, began by believing in the theory of brain plasticity but found that his own work led him back to reexamine the issues of localization and connection in the brain.[39] The experiments in question involved the severing of the corpus callosum, the wide, flat bundle of neural fibers connecting right and left hemispheres. Though early studies had regarded it as useless, by the 1920s some neurologists speculated that the corpus callosum was "the principal organ of the Self" because it "lies equally on one and the other upper rim of the two cerebral hemispheres," in its uniqueness recalling Descartes's arguments for the special significance of the pineal gland as the meeting point between mind and body.[40]

Sperry and his colleagues surgically severed the corpus callosum and the optic chiasma in a cat so that the information from its left eye went solely to its right hemisphere, vice versa for the other eye and hemisphere. Though in most situations the cat behaved normally, if one eye were covered, each hemisphere seemed to have to learn separately what to do. The cat seemed to have two separate minds. In the 1960s, similar procedures were done to try to relieve human patients suffering from intractable epilepsy; the results were similarly startling. As Sperry described it, when faced with tests of the separate sides of the brain, "instead of the normally unified single stream of consciousness, these patients behave in many ways as if they have two independent streams of conscious awareness, one in each hemisphere, each of which is cut off from and out of contact with the mental experiences of the other."[41]

Interest in multiple personalities continued to grow during the twentieth century, as clinicians brought forward more and more extreme cases involving dozens of selves (or even more) within a single individual. By the 1980s, the matter had become extremely controversial in light of the possibility that "false memories" (perhaps even implanted during therapy) could then be misinterpreted as evidence of another autonomous self. In 1994, the official psychiatric diagnosis of "multiple personality disorder" was reconstituted as "dissociative identity disorder," its change of name indicating the continued deep controversies about the true nature and significance of these manifestations.[42]

Though these clinical investigations drew attention to possible schisms in the self, Max Weber emphasized the integrative powers of polyphony in his pioneering account of the sociology of music. Weber's unfinished work, *The Rational and Social Foundations of Music* (1911), formed part of his larger enterprise to describe the ways that society gradually imposed order on the human world through the process of "rationalization." By this term, Weber seems to have meant a structure of rules calibrated through reason to achieve certain ends; for instance, in rational bureaucracy the rule of law replaced traditional ways of doing things or the individual whim of king or lord. Though he noted the ways in which rationalization made possible greater order and equality, Weber also recognized that it "disenchanted the world," replacing the mysterious workings of inspiration and tradition with rational calculation. He was quite aware of the costs and sacrifices involved in that disenchantment, as well as what its value might be. As Joachim Radkau summarized Weber's views, "the ear of modern Western man has been aligned and deadened by the organ and piano, machine-like instruments with major chords. ... The progress of instrumental technique does not bring about a higher development of human sense-perception—quite the contrary." Though musical instruments were a "crutch," through them Weber thought we could achieve "great steps" beyond what the unaided human voice could do.[43]

For Weber, music was a particularly clear example of the ways that the rules of counterpoint put in order what had previously been intuitive and unstructured styles of composition, exchanging the magic of inspiration or unaccountable virtuosity for rational procedures. Deeply influenced by Helmholtz's work (though at times critical of it), Weber also used the growing body of studies of non-Occidental music to illuminate what he thought was the special history of Western music.[44] Using German terms that make finer distinctions than our more inclusive English words, Weber set up a spectrum between three "pure types." By "many-tonedness" (*Mehrtönigkeit*) or "many-voicedness" (*Mehrstimmigkeit*) he meant the simultaneous sounding of different notes such as he knew from Greek or Oriental music, which "bear no relation to our harmonic progressions" and aim only at "fullness of sound."[45] "Harmonic-homophonic" (*harmonisch-homophone*) music "represents the subordination of the entire musical sound formation [*Tongebildes*] under one voice carrying the melody."[46] Finally, polyphony proper (*Polyphonie*) has "several voices of equal standing run side by side, harmonically linked in such a way that the progression of each voice is accommodated to the progression of the other and is, thus, subject to certain rules."[47]

Thus, like the citizens of a modern rationalized state, these polyphonic voices claim equal rights and treatment under the law: "In modern polyphony these rules are partly those of chordal harmony, partly they have the artistic purpose of bringing about such a progression of voices that each single one can independently come to its melodic right. Still, and possibly because of it, the ensemble as such preserves a strict musical (tonal) uniformity."[48] Weber noted that "polyphony in this sense, especially contrapuntalism [*Kontrapunktik*], has been known in the West in highly developed conscious form only since the fifteenth century. It attained its highest perfection in the work of Bach."[49] For Weber, the triumph of polyphony represented and reflected the perfection of rational order in human society.[50]

Writing during the same years before World War I, Marcel Proust tended to emphasize the dissonances within and between his characters, their unstable and divided souls, rather than Weber's vision of rationality and concord. Drawing on music for some of his most vivid descriptions of states of love and obsession, in Proust's *Search for Lost Time* the narrator learns to go beyond his naïve belief that "it was through words that the truth was communicated to other people."[51] Instead, he realizes the power of sounds, "which have no fixed point in space."[52] Proust draws attention to the strange "jumble of sound which, in Molière's comedies, is produced by several people saying different things at one and the same time"; likewise, Proust describes two of his own characters, each of them "monologists, to the extent of being able to bear no interruption," speaking simultaneously at cross-purposes.[53] Often Proust uses the piano to suggest the polymorphous possibilities of music, which is "not a miserable scale of seven notes, but an immeasurable keyboard still almost entirely unknown on which, here and there only, separated by shadows thick and unexplored, a few of the millions of keys of tenderness, of passion, of courage, of serenity which compose it, each as different from the others as one universe from another, have been found by a few great artists who do us the service, by awakening in us something corresponding to the theme they have discovered, of showing us what richness, what variety, is hidden unbeknownst to us."[54]

Using music to explore "that great unpenetrated and disheartening darkness of our soul which we take for emptiness and nothingness," Proust explores the multiple selves within a single character, who may on some occasions be capable of great kindness, while on others seem cruel; such seeming contradictions appear in the seemingly straightforward Françoise (the narrator's beloved family servant) no less than in the enigmatic Baron de Charlus.[55] A person's divergent behaviors in sufficiently different contexts may reveal the lack of any such underlying, fixed character, to the extent that Proust at times will refer to certain momentary aspects of his characters as a more accurate way of describing how they seem to be different persons on different occasions. Thus, he chose to speak specifically of "the Swann of Buckingham Palace" rather than pretending that Charles Swann was a single consistent being at all times; likewise, "Albertine in her rubber raincoat" was a different person than Albertine at other times.[56]

Nor did Proust treat this multiplicity of selves as merely a perspectival experience of others as we observe them. As his narrator awakened in a strange bedroom, he realized the perils involved in recovering the self each morning:

But it was enough if, in my own bed, my sleep was deep and allowed my mind to relax entirely; then it would let go of the map of the place where I had fallen asleep and, when I woke in the middle of the night, since I did not know where I was, I did not even understand in the first moment who I was: I had only, in its original simplicity, the sense of existence as it may quiver in the depths of an animal; I was more destitute than a cave dweller; but then the memory—not yet of the place where I was, but of several of those where I had lived and where I might have been—would come to me like help from on high to pull me out of the void from which I could not have got out on my own; I crossed centuries of civilization in one second, and the image confusedly glimpsed of oil lamps, then of wing-collar shirts, gradually recomposed my self's original features.[57]

To do so, the narrator needed to pass through "sometimes one, sometimes another, of the bedrooms I had inhabited in my life, and in the end I would recall them all in the long reveries that followed my waking."[58] In this process, he only gradually recomposed his "original features" by reference to those remembered bedrooms and the various recollected positions of his body in their beds. Here, the polyphony of inner selves emerges through their external correlates, the rooms that complement and compose the different selves "he" was when sleeping in them. In the course of awakening, the rooms and the selves initially cannot be distinguished; gradually, they help to recover and connect each other, finally restoring the sleeper to his most recent room and self. Proust notably does not insist on any single one of those selves as uniquely or exclusively authentic; he leaves us contemplating their multiplicity as his dreamer experiences their polyphonic unfolding in recollection and reverie.

Telling the story of Charles Swann, Proust further acknowledges that someone (through love and suffering) may change so deeply that he or she no longer is the same person: "For one cannot change, that is to say become another person, while continuing to acquiesce in the feelings of the person one has ceased to be."[59] Here the alteration is so radical that it is no longer a question of which self, out of many, has momentarily come to the surface but rather which self has died in giving life to another. These selves, dying and newly born, give life and poignancy to Proust's story, asking us into what new self might we yet find ourselves transformed. Perhaps mercifully, a certain obliviousness shields the new person from a real awareness of his or her old, dead self, for "it is so difficult to duplicate oneself and give oneself a truthful display of a feeling one no longer has."[60] But other people are able to see the difference, to compare the old and new selves, as the narrator does in telling the story of Swann's love and how it changes him. Such an external observer can more securely register the paradox of a new self inhabiting the old body than can the changed self, already forgetful of who it once had been. Divining Swann's inner life, the narrator notices that, "like certain novelists, he had divided his personality between two characters, … he reproduced himself by simple division like certain lower organisms."[61]

Other writers and theorists explicitly enlisted the concept of polyphony as illuminating their practice. Beginning about 1914, Amy Lowell advanced a concept of "polyphonic prose" as "an orchestral effect" that would treat subjects "at once musically, dramatically, lyrically, and pictorially."[62] She considered this new kind of prose as reflecting modernity, her work forming part of the imagist movement along with Ezra Pound, who also referred to polyphony as a poetic desideratum.[63] Lowell especially applied this technique to descriptions of the modern city; by polyphony she meant a sonorous evocation that merged different senses, a kind of artistic synesthesia. For instance, in her 1916 poem "Spring Day," she connected color, shape, and sound by describing how "a stack of butter-pats, pyramidal, shout orange through the white, scream, flutter, call: 'Yellow! Yellow! Yellow!'"[64] Her emphasis seemed more on the sheer multiplicity of appeals to different senses than on the simultaneity of different experiences within a single sense that seems so important to musical polyphony.

In his theory of literature emphasizing its possibilities as an unfinishable conversation between different voices, Mikhail Bakhtin described some works as particularly "polyphonic," especially Fyodor Dostoyevsky's novels in which several diverse voices speak: "What unfolds in his works is not a multitude of characters and fates in a single objective world, illuminated by a single authorial consciousness; rather a *plurality of conscious nesses, with equal rights and each with its own world*, combine but are not merged in the unity of the event."[65] In contrast, Leo Tolstoy's writing "contains neither polyphony nor (in our sense) counterpoint. It contains only one cognitive subject, all else being merely objects of its cognition." This is "the author's world, an objective world vis-à-vis the consciousnesses of the characters. Everything within it is seen and portrayed in the author's all-encompassing and omniscient field of vision."[66]

As powerful as Bakhtin's insight may be about these two contrasting modes of literary creation, however, we should note his very particular use of "polyphony." As different as Alyosha Karamazov may be from his brothers Ivan or Dmitri in character and "voice," they never speak simultaneously but sequentially—nor could they possibly speak simultaneously, in prose narration as we know it. By emphasizing what he calls the "dialogic" quality of their encounter, Bakhtin praises the way one voice can respond to another, sometimes deeply changing as a result of their dialogue, which must therefore come in sequence, one speaking after hearing the other. Having both speak at once would suggest, on the contrary, a lack of mutual attention or listening, each merely blurting out what they had to say, heedless of the other.

Still, we have some important precedent for considering that "polyphony" could be implicit even in a single melodic line, as in the example given above of the solo cello fugue in Bach's Fifth Suite: a single voice suggested the dialogue of independent lines, as the fugue emerged (which we called "virtual polyphony"). Yet Bakhtin's notion of polyphony seems to strain against this reading, for he treats Dostoyevsky's characters as interacting far more conflictually than do the voices in the Bach cello fugue, which seem integral parts of a single melodic line. If indeed Alyosha and Ivan are entirely separate, then the silences of one cannot simply be read as harmonizing with the other, for their views—and their basic individualities—seem much less in accord: they "combine but are not merged." Nonetheless, Bakhtin's larger point is that Dostoyevsky's characters speak with such a degree of independence—and apart from the controlling direction of an authorial voice—that their intercourse seems to him more like polyphony than monophony. Bakhtin's praise of a novelistic "plurality of consciousnesses" implies his preference for the manyness of polyphonic voices as having more life than a purely author-dominated text. Aware of the artifice involved in authors appearing to erase their own dominance over their characters, Bakhtin draws attention to the felt power of characters that strain against each other, thereby enacting such independence and freedom that ultimately may render a deeper truth than can be presented by only one voice.

Considered as polyphony, Dostoyevsky's novels are markedly "dissonant," a quality he uses to great dramatic effect: the push-pull between his characters, the love-hate, the sheer intensity of their interactions gives a special feeling to his writing. That same musical element underlies the concept of cognitive dissonance introduced by Leon Festinger in social psychology, meaning the mental stress experienced by those who hold contradictory beliefs or who perform actions or encounter information contradictory to their beliefs. This concept implicitly analogizes mental states to the relative feeling induced by hearing polyphonic intervals: the greater or lesser tension felt between simultaneous voices. Though Festinger considered other possible terms such as "inconsistency" or "discomfort" (with their more unequivocal sense of something entirely to be eliminated), ultimately he settled on "dissonance," with all its musical connotations that indicate a more complex state of arousal that seeks a lowering or resolution of dissonance, rather than its mere elimination.[67] In particular, social psychologists have conducted experiments specifically designed to distinguish conflict from dissonance.[68]

Cognitive dissonance has become a well-established psychological concept with a wide range of applications, from individual to group dynamics. For instance, in his study of cults that predicted the end of the world, Festinger showed that even a seemingly crushing disconfirmation (such as passing the "absolutely certain" date predicted for Armageddon) would not defeat the cult. Paradoxically, its more zealous members would then begin to proselytize even more ardently, which Festinger explained as their attempt to lower the felt dissonance between the failed prophecy and their fervent convictions.[69]

Similar issues of dissonance are also important in sociology, in the context of the relative coherence or incoherence of social groups in terms of an understanding of how those qualities emerge from the individuals who constitute the group. On the one hand, Pierre Bourdieu emphasized the essential coherence of the "habitus" of those individuals, meaning "a permanent and general disposition towards the world and others," the ways in which their beliefs and practices habitually cohere.[70] For instance, "an old cabinetmaker's world view, the way he manages his budget, his time or his body, his use of language and choice of clothing are fully present in his ethic of scrupulous, impeccable craftsmanship and in the aesthetic of work for work's sake which leads him to measure the beauty of his products by the care and patience that have gone into them."[71]

On the other hand, such a unified personal identity may be a kind of social illusion. Erving Goffman argued that, though we think we can gather who a person "really is" from his behavior in one situation, "that is no reason to think that all these gleanings about himself that an individual makes available, all those pointings from his current situation to the way he is in his other occasions, have anything very much in common."[72] Thus, we find situations and novels engrossing precisely because at every moment we may find in them evidence of new and surprising sides of an individual that we had not known from other situations involving that same character.

Bernard Lahire emphasized the social context of these seeming disjunctions: because we occupy different social positions in different situations (variously "child," "lover," "parent," "colleague," "pitcher," "employee"), "we live experiences that are varied, different and sometimes contradictory. A plural actor is thus the product of an—often precocious—experience of socialization in the course of their trajectory, or simultaneously in the course of the same period of time in a number of social worlds and occupying different positions."[73] Among philosophers, David Hume famously argued that the appearance of having a unified personal identity was an illusion, based on a bundle of qualities we endow with enduring selfhood merely out of habit. For Lahire, these qualities are not just an arbitrary stack or heap because they turn out "to be organized in the form of social repertoires. … People learn and understand that what is said and done in one context is not said or done in another."[71] They accordingly develop a sense of what behavior would be consonant in that social context, what is therefore expected of them, and act accordingly; through long habit, they sense departures from those expectations acutely and immediately, seeking to diminish any such dissonances as if they were perfectly ingrained in their individual "character."

As these examples show, the concept of polyphony—even when not explicitly invoked or mentioned—helps us understand these wide-ranging developments across many fields, which all concern the interrelation of many participants within a larger whole. Because of this underlying basic problem, the spectrum of polyphonic experience—from the most unified and consonant to the most dissonant or heterogeneous—offers a common framework and vocabulary whose generality can include the multiple selves of literature or psychology as well as the social matrix in which they all struggle to cohere.

14 Tuning the Brain

Though many aspects of its functioning remain unclear, there is broad and long-standing consensus that the brain has no single "command center" but operates through the flexible interaction of many neural circuits.[1] My final chapters will consider this central fact through the concept of polyphonicity. During the twentieth century, the growing understanding of neurons and their circuits provided the physiological basis for an understanding of mental activities as emerging from their interaction; the discovery of electroencephalography (EEG) gave access to overall brain states. Computers developed sequential processing to mimic the parallel processing of the brain, and artificial networks offered models of neural circuits; despite their differences, both natural and artificial neural networks are describable through concepts of reverberation and dissonance. In general, the "polyphony" of neural circuits may regulate itself through dissonances among them without any central control, giving rise to more or less coherence as the whole system autonomously finds states of relatively lower dissonance (and energy) overall. In turn, characteristic features of dissonance and its resolution may provide a general understanding of seemingly paradoxical features of learning; a different relation to cerebral polyphony may also illuminate mental conditions including ADHD (attention-deficit/hyperactivity disorder). Throughout this chapter and the next, polyphony will emerge as an analogy used by neuroscientists to describe their findings. At the end of this chapter, we will pause to assess the meaning and significance of this analogy.

The most fundamental element of the contemporary consensus about the brain is that it contains no homunculus, no "little person" responsible for consciousness, somehow beholding in an inner "Cartesian Theater" the inputs of the senses and putting them together.[2] Indeed, even before Descartes, the very idea could well have seemed logically preposterous: if my consciousness is caused by a little person inside me, must not there then be a still smaller person inside him, and so on ad infinitum? Though lacking any single control center, the brain does show certain kinds of hierarchical structure. For instance, investigations of the visual system have delineated a hierarchy of levels of processing of the raw input from the retina via the optic nerves. Thus, if V1, the primary visual cortex, is damaged, a person may still receive color information via the retina but will not be able to process it into color vision, only black and white.[3] In many other cases, it seems that sensory nerve inputs merge into various higher-level centers that then feed into yet higher centers.

Absent a single "top-down" center, the remaining alternative appears to be that the interaction of many cerebral subcenters gives rise to what we term sensation, awareness, and consciousness through a kind of polyphony of those centers, a complex interweaving of relative dissonance or consonance between them, whose felt correlates are the facts of our mental life. Of course, the nervous system and brain operate through electrochemical signaling, not sound, but polyphony may be a helpful analogy by which to understand mental functioning. The question then becomes how neural structures could exhibit such polyphonicity.

Using Camille Golgi's method of staining a few nerve cells, in 1888 Santiago Rámon y Cajal presented microscopic evidence that the nervous system of birds was composed of discontinuous nerve cells that came to be called *neurons*.[4] Since then, more and more clarity and detail has emerged about the functioning of individual neurons and the synapses that connect them, the all-important gap or "cleft" where the axon of one neuron communicates with the dendrite of another (figure 14.1).[5] Though neurons generally obey an "all-or-nothing" principle (either an individual neuron fires or it does not, nothing in between), greater intensity of stimulation can alter the frequency of the transmitted pulse, though not its amplitude.[6] Thus, we can simplify the complex interactions of a group of neurons by focusing attention on the frequencies of the pulses they exchange among them. As we will discuss further in the next chapter, each neuron is not just a simple on-off switch but is also capable of more complex calculations involving (for instance) the relative amounts of excitation and inhibition it receives from other neurons around it. This detailed "wiring" is essential to the exact character of each neuronal circuit and therefore to its ability to respond to changing circumstances of external stimulation (for instance). Because of this wiring, an assembly of neurons can construct a feedback loop or many other kinds of adjustable circuits that can respond to changing stimuli.

Oversimplifying these manifold complex possibilities, we can concentrate on the frequencies of the interchanged signals first between a group of individual neurons forming a simple circuit, then between that circuit and others in the brain. Neuroscience tries to understand the neurons, their circuits, and the general network they all comprise, though each part is itself already very complex. Indeed, Rafael Yuste has emphasized the fundamental importance of the shift from the "neuron doctrine" to neural networks as "ensembles that generate emergent functional states that, by definition, cannot be identified by studying one neuron at a time."[7]

As early as the 1930s, Cajal's student Rafael Lorente de Nó argued that the structural design of the nervous system involved recurrent interconnectivity that could generate "functional reverberations," meaning patterns of neuronal activity that persisted after the initial stimulus had ceased.[8] Building on Lorente de Nó's work, in 1949 the psychologist Donald O. Hebb "was among the first thinkers who explicitly stated that the brain's ability to

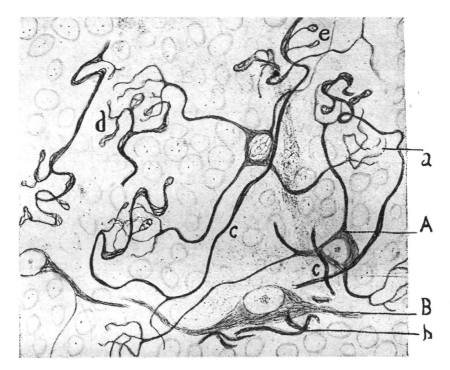

Figure 14.1
Santiago Rámon y Cajal's drawing of axons (*c*, *d*, *e*) and dendrites (*a*) in the cerebellum of an adult cat.

generate a coherent thought derives from the spatiotemporal orchestration of its neurons," his "cell assembly" hypothesis, as György Buzsáki describes it.[9] To underlie these interactions, Hebb proposed that, between neurons, "the persistence or repetition of a reverberatory activity (or 'trace') tends to induce lasting cellular changes that add to its stability," which he phrased as a fundamental assumption: "*When an axon of cell A is near enough to excite a cell B and repeatedly or persistently takes part in firing it, some growth process or metabolic change takes place in one or both cells such that A's efficiency, as one of the cells firing B, is increased.*"[10] This "synaptic plasticity rule" gives concrete physiological form to a process of what came to be called *Hebbian learning*, emerging from the networks of neurons themselves, through which "information reverberates within the formed assembly … in loops of synaptically connected chains of neurons. This reverberation explains why activity can outlast the physical presence of an input signal," as Buzsáki puts it.[11] Further, these assemblies are not fixed groups of neurons but flexible: "Hebb's cell assembly is a transient coalition of neurons, much like the dynamic interactions among jazz musicians."[12] The neurons' reverberation (another acoustical term) underlies learning, in Hebb's view. He coined the name *connectionism* to describe his view that interneural "connections rigidly determine what [the] animal or human being does, and their acquisition constitutes learning," as opposed to "field theory," which "denies that learning depends on connection at all" and treats the cortex "as a statistically homogeneous medium."[13]

These reverberating neural networks are "polyphonic" by virtue of the *multiplicity* of their constituents and because they affect each other through their *frequencies*, analogous to the pitches of an unimaginably complex polyphonic composition.[14] To be sure, there is no composer orchestrating this cerebral polyphony; its "voices" proceed without any prior control, though they interact constantly via the mutual reception and interplay of neuronal frequencies, thereby exciting and inhibiting each other ceaselessly. Perhaps the best example we have seen might be Kepler's cosmic motet, in which he understood the individual voices of the planets as being more or less dissonant with each other and forming a seven-voice composition that constantly tracks these relative dissonances as each planet follows its own orbit around the sun.

For Kepler, the prevalent dissonance that mostly characterizes the "harmony of the world" was responsible for the "misery and famine" all around us. Still, the degree of dissonance could increase or decrease as the planets moved in their orbits. What if we use Kepler's approach and treat the brain as an even more complex "motet" comprising all the interplay of the neurons collectively? At first glance, it seems that the resulting "music of the hemispheres" would be intolerably dissonant and chaotic, yielding nothing that could correspond to consciousness, attention, or thought. If Kepler's seven planets already were so dissonant, would not the 10^{11} neurons in a human brain produce nothing more than cacophony or just "white noise," an equal mixture of all frequencies?

In fact, these complex polyphonic frequencies within the brain can sum to surprisingly simple patterns. In 1875, the Liverpool physician Richard Caton applied a galvanometer to the external surfaces of rabbit and monkey brains and noted that they were "usually positive in relation to the surface of a section through" the brain, showing "feeble currents of varying direction." Moreover, "the electric currents of the grey matter appear to have a relation to its function. When any part of the grey matter is in a state of functional activity, its electric current usually exhibits negative variation. … Impressions through the senses were found to influence the currents of certain areas," so that if the retina were stimulated by light, currents in the corresponding area of the brain (in the opposite hemisphere) would be "markedly influenced."[15]

The currents and voltages involved, however, were so small that it required sensitive instruments to record them precisely. In 1924, Hans Berger recorded the first human electroencephalogram (EEG) using electrodes attached to the skull.[16] He named the first rhythmic EEG activity he observed the "posterior basic rhythm," which he later called the "alpha wave" (figure 14.2). This rhythm seemed to characterize the brain activity in the posterior part (and stronger in the dominant hemisphere) of the brains of relaxed subjects whose eyes were closed; the alpha rhythm would attenuate if they opened their eyes or exerted their minds in other ways. The alpha rhythm has a characteristic frequency range

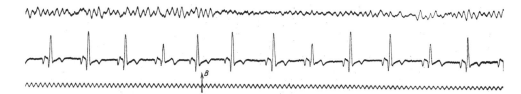

Figure 14.2
Hans Berger's recording of an EEG (1930), showing at top (from left to right) the alpha rhythm of a resting subject, a thirty-year-old physician, "Dr. V"; at *B*, the back of his right hand began to be touched and stroked with a glass rod, at which point the EEG rhythm became more rapid and smaller in amplitude (beta rhythm). Gradually thereafter, the alpha rhythm was reestablished. In the middle, the electrocardiogram shows a steady rhythm. At the bottom, time in intervals of 0.1 seconds (a frequency of 10 hertz).

of 7–14 hertz; in his *Music for Solo Performer* (1965), Alvin Lucier amplified the performer's brain waves to play various percussion instruments, which would rattle eerily when an alpha rhythm predominated.[17] As the brain becomes more active, higher frequencies emerge in the EEG; normal waking consciousness seems related to the beta waves (whose frequencies are 15–30 hertz). More recently, digital EEGs allow the observation of gamma waves (30–100 hertz, typically near 40) that arise while performing cognitive tasks.[18] Besides these, sleeping adults manifest the very slow (but high amplitude) delta rhythm (4 hertz).

The various rhythms' sequence of frequencies suggests a simple progression from deep sleep (delta) through successively more active mental and physical states (alpha, beta, and gamma). Thus, the brain's tremendously complex networks of neurons can indeed collectively produce certain simple electrical rhythms that are the overall result of their intricate polyphony. Though these rhythms are electrical, one may well wonder why they fall into the frequency ranges observed, rather than being many times smaller or larger; by comparison, visible light spans the much higher frequency range of $4–8 \times 10^{14}$ hertz. It seems scarcely coincidental that the cerebral frequency bands reflect the lower part of the human perceptive range, which can sense 4–16 hertz via touch, 12 hertz being the lowest audible musical tone (under ideal conditions) and 20–20,000 hertz the nominal range of human hearing.

The most alarming sounds in the environment of early humans would arguably have been those made by large predators; for instance, the lowest frequencies of a lion's growls are around 10 hertz, near the bottom of the human hearing range (though the high notes of its roars can reach 430 hertz).[19] The frequencies of wakefulness and then concentration or action (beta and gamma) span this range of dangerous roars so that humans have the capacity to respond to an alarming stimulus in those frequencies. Friedrich Nietzsche called the ear "the organ of fear" and thought that, more than any other sense, hearing "could have evolved as greatly as it has only in the midst of obscure caves and woods"; if so, the frequency response of the brain would face strong selective pressure to recognize those menacing auditory frequencies.[20] Of course, the auditory frequency range may have also been set by the frequency response of the neurons, limited by their electrical and chemical processes.

Building on his conception of cybernetics, in 1961 the mathematician Norbert Wiener suggested "the possibility that the brain contains a number of oscillators of frequency nearly 10 per second, and that within limitations these oscillators can be attracted to one another. Under such circumstances, the frequencies are likely to be pulled together in one or more clumps, at least in certain regions of the spectrum."[21] By this general process of "autocorrelation," oscillators (such as pendulums or other periodic phenomena) with some amount of coupling tend to synchronize ("attract" each other, in Wiener's terminology).[22] For instance, someone on a playground swing will tend to excite sympathetic oscillations of nearby swings via the coupling of their shared suspension (♪ sound example 14.1). This is the generalization of long-known phenomena of sympathetic vibration or resonance: singing a pitch into a piano (with the pedal down) excites its strings having that frequency (or closely related overtones, such as the fifth or octave). Wiener's suggestion offers a general path to understanding how the observed bands (which he calls "clumps") of brain rhythms could emerge from the welter of underlying frequencies that make up the cerebral polyphony.

Others used these concepts to address the fundamental *binding problem* of neuroscience: how can the brain unite its separate perceptions of the color, shape, movement, and sound of an object so that we perceive a *single* thing, not a disconnected assortment of felt qualities? To do so, the neurons involved might fire synchronously, their coordinated activity causing the oscillations manifest as wave patterns in the EEG. Thus, in 1989 Wolf Singer and his collaborators used their observations of gamma oscillations (40–60 hertz) in the cat visual cortex to argue that "synchronization of oscillatory responses in spatially separated parts of the cortex may be used to establish a transient relationship between common but spatially distributed features of a pattern," essentially the binding between such disparate features as the orientation and motion of a figure.[23] Supporting Singer's idea, Francis Crick and Christof Koch conjectured that gamma oscillations "*might be the neural correlate of visual awareness*" and emphasized the special activity of the thalamus, a region located deep within the brain whose multiplicity of connections to the cortex seems to result in gamma oscillations.[24] They compared the thalamus to the "conductor" of the cerebral music, though noting that "awareness requires the activity of the various cortical areas as well as the thalamus, just as a conductor needs the orchestra to produce music."[25] Whether "conducted" by the thalamus or not, their interpretation relied on the polyphonic (even orchestral) aspects of the mind.

Going even further, Rodolfo Llinás argued that the thalamocortical system "synchronously relates the sensory-referred properties of the external world to internally generated motivations and memories. *This temporally coherent event that binds, in the time domain, the fractured components of external and internal reality into a single construct is what we call the 'self.'*"[26] For Llinás, "self" does not mean even the thalamus as the cerebral orchestra's conductor but is a "calculated entity" he considers to be "a convenient word that stands for as global an event as does the concept of Uncle Sam *vis-à-vis* the reality of a complex, heterogeneous United States."[27]

If so, the many streams of processing within the brain may connect into something like a single stream. Here we confront the difference between the brain and a computer that is ultimately a Turing machine, an idealized device capable of executing a sequence of computations that Alan Turing first imagined in 1936 as unrolling on a tape in linear sequence.[28] By 1956, John von Neumann (who made important contributions to the practical implementation of Turing's ideas in the first computers) noted that

an efficiently organized large natural automaton (like the human nervous system) will tend to pick up as many logical (or informational) items as possible simultaneously, and process them simultaneously, while an efficiently organized large artificial automaton (like a large modern computing machine) will be more likely to do things successively—one thing at a time, or at any rate not so many things at a time. That is, large and efficient natural automata are likely to be highly *parallel*, while large and efficient artificial automata will tend to be less so, and rather to be *serial*.[29]

As von Neumann anticipated, the brain indeed has many neural subcenters operating in parallel. Their interconnection allows these multiple centers to act in concert and give the overall semblance of a coherent state such as would arise from the serial processing envisaged by Turing and von Neumann. In this way, parallel processing can overcome the limitations of serial processing; because of this, a single human brain has more computing power than all the computers on Earth, as of this writing. To revert to a sonic analogy, an entire orchestra producing hundreds of simultaneous sounds sums into a single complex wave that can shape the groove of a record or be encoded as digital information. Likewise, the innumerable signals within the brain sum to a single (if complex) wave; Llinás thinks of the brain not as "a slumbering machine to be awoken by the entry of sensory information, but rather as a continuously humming brain."[30]

Beginning in the 1990s, the study of artificial neural networks offered simplified models of neural circuits that exhibited self-generated coherence and even learning, in the sense that Hebb introduced. These networks can be represented using computer code or increasingly by specially designed chips. Such approaches recognize that the brain is not a computer in the ordinary sense, executing a single preprogrammed series of operations, but instead operates through *parallel distributed processing* (PDP): parallel in the sense of many neural circuits active simultaneously, distributed throughout the brain rather than localized in any one area.[31] This basic parallelism expresses one aspect of polyphony, the multiplicity of simultaneous voices. These networks also embody the other aspect of polyphony in the way the nodes in the network are interconnected to register their relative states, whether more or less aligned and hence dissonant. Further, these internodal dissonances are not fixed or static but continually changing, just as the relation between different polyphonic voices constantly changes, becoming more or less dissonant or consonant from moment to moment.

These dynamic dissonances reflect the underlying physiology of the neuron, which can receive input from many different synapses (connected in turn to other neurons). Each neuron will or will not fire based on whether a weighted sum of the inputs it receives from other neurons exceeds a certain threshold. Physiologically, the weights depend on the strength of the synaptic transmission and thus are variable, so that the network can respond to new situations, not merely repeat the same pattern of firing without changing. Because these weights can change in response to stimuli, the whole network can adapt and even learn. Then, imitating these principles, artificial neural networks currently perform many tasks that ordinarily would require human intelligence, including facial recognition or the recognition of complex patterns (such as classifying galaxy shapes or tracks in accelerator experiments). Here again, the network does not merely register patterns from a preprogrammed memory, which would restrict it to its initial data. Rather, the possibility of changing weights allows the network actually to *learn*: a function built into the network gauges how closely its output (say its categories of distinct facial types) matches the input it received (a training set of individual faces), then gradually adjusts the weights so as to maximize this matching. My son's smartphone catalogs his photos in ways he never programmed. From the images he has taken, it learns salient features it uses to create new categories (such as "pictures of ramen") and to recognize and remember faces. Even if we stop short of the more extreme claims such as Llinás made, it seems reasonable that any model capable of learning, as artificial networks do, has captured some essential feature of the mind, understandable in terms of the polyphonic resolution of dissonance.

Rather than preset rules, such a network adjusts the variable weights at its nodes through applying a function that gauges how closely the net's output matches the input it received. In one of the more prominent versions of these networks, this function is even called the *harmony function*, making quite explicit that the criterion by which the net incrementally improves its performance is the "harmony"—the relative consonance—between its results and the data it aims to match.[32] Often the network model builders will speak of how that harmony function "tunes" the weights to minimize the net dissonance of the net's results.

The proponents of artificial neural networks thought that they were realizing Hebb's "switchboard theory" of the brain and thus called themselves "connectionists."[33] Yet though such networks generated much attention during the 1980s, some neuroscientists objected that these artificial networks could not represent the actual functioning of the brain. In 1989, Crick noted that neurons do not transfer information backward, as the "back-propagation" used in artificial neural networks would require. He also criticized these models as "cavalier" in their approach to the actual diversity and characteristics of neurons.[34]

Stepping back from the problems of artificial neural networks, let us consider a general concept of "tuning" that may apply to them as well as to the actual functioning of the brain. Tuning describes an exquisitely sensitive adjustment such as singers would apply in order to control their relative dissonances and consonances as clearly and intentionally as possible. Anyone who has ever tuned a string instrument knows the experience: turning the peg slowly, approaching the desired pitch but generally overshooting, then undershooting, moving back and forth while gradually settling into the "center" of the pitch. Singers experience a similar process but usually more fluidly (lacking any external peg to monitor) as they adjust their pitches so as to control the dissonances, centering their own sound and balancing with the other voices.

With this in mind, we can further refine the felt experience of dissonance as it emerges in music and thereby also informs the comparison with neuronal dissonance. Let us return to Helmholtz's curve showing the relative roughness of two simultaneous notes on a violin (figure 11.8a, shown again as figure 14.3), which we earlier considered as a representation of the primordial musical consonances emerging from a continuum of intervals. Now let us examine in more detail how these consonances emerge.

Helmholtz plotted the relative roughness caused by the beats between two notes less than an octave apart (from c′ to c″); as noted above, the well-known consonant intervals (such as the octave c″, the fifth g′, the major third e′) stand out as the lowest points in that curve. When we ask how those relative consonances are approached from the notes immediately on either side of them, we might have intuitively expected that dissonance would simply decrease as a consonant interval is approached. On the contrary, in each case as we approach a consonance, the degree of dissonance *rises* noticeably before it finally drops down to a relative minimum at the consonance in question. This is true whether we slide to that consonance from just below or just above it. Furthermore, the relative degree

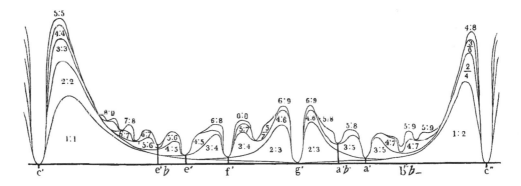

Figure 14.3
Hermann von Helmholtz's diagram of the relative degree of dissonance between two notes within an octave (from c′ to c″), plotted against their relative distance. The consonant intervals are listed with their note names. As each of them is approached (either from above or below), the degree of dissonance markedly increases before it decreases.

to which this happens corresponds directly to the perceived degree of consonance that is approached: for instance, in figure 14.3 when approaching the octave (c″), the peak of dissonance is much higher than the corresponding peak approaching the fifth (g′), which in turn is higher than the peak near the major third (e′). We infer that the relative degree of consonance is therefore also directly correlated to the corresponding degree of dissonance as one approaches it.

Helmholtz's diagram came from his calculations of the relative degree of beating (interference) between the overtones of two notes that are allowed to slide continuously over an octave. Experimental studies with human subjects also reveal the same basic finding that dissonance increases as two pitches approach a consonant interval.[35] Setting aside the subtleties due to the varying bandwidth of intervals (the degree to which different intervals between pitches can be distinguished by the vibrating basilar membrane of the inner ear), Helmholtz's explanation relies on the mathematics of combining waves, in which the felt sensation of "dissonance" is associated with the degree of beating between the waves. The ear and brain then convert that increased beating into the felt sense of greater roughness or dissonance, which peaks just before its sudden drop as the consonant interval is reached.

Helmholtz's discovery of the sharp rise of dissonance in the neighborhood of consonance helps explain the approach to relatively consonant states throughout the brain. If indeed the brain is a polyphonic network involving relative states of dissonance between its nodes, then Helmholtz's observation implies that states of consonance in brain function are generally approached through heightened transient dissonances. Based on the general concept of dissonance, we would accordingly expect that the "consonant" state in which the brain has settled down into a stable recognition of some concept would be preceded by a sharp *increase* in the failure rate, rather than a uniform decrease.

This helps explain the otherwise paradoxical phenomenon of U-shaped learning curves. Consider the way children learn to solve such "equivalence problems" as $7 + 4 + 5 = 7 +$ ___. One might expect intuitively that the ability to solve such problems would increase steadily with age, but in fact observational data show otherwise (figure 14.4): seven-year-olds routinely outperform eight-year-olds. Only well into their ninth year do children finally exceed the abilities to solve these problems they showed at seven.[36]

One can interpret the U-shaped dip in these learning curves as an effect of cognitive dissonance within the minds of the individual students as they try to solve their math problems. As with Helmholtz's curves, the shape of the dip reflects the precise structure of dissonance and resolution. If the general concept of dissonance applies to the brain, learning proceeds not uniformly or directly but involves an essential phase of markedly increased difficulty and failure that precedes the "breakthrough" to a new cognitive synthesis: things get worse before they get better, as therapists often remind their patients and teachers their students. Indeed, the dissonance that surrounds learning increases just

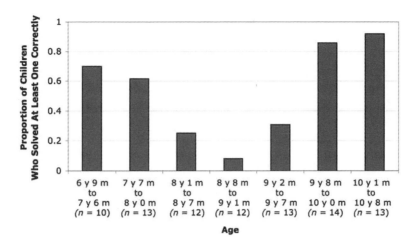

Figure 14.4
Proportion of children who solved at least one equivalence problem correctly (from McNeil 2007), their age given in years (*y*) and months (*m*). Note that seven-year-olds outperform eight-year-olds, shown in the U-shaped curve of the results.

as the learner is getting closer and is more noticeable in proportion to the difficulty of the material. Only after learners have gone through this transient but quite trying cognitive dissonance many times do they become familiar with the process and thus less averse to the pain.[37]

When social psychologists took up the concept of cognitive dissonance, they treated it as expressing conflicts between beliefs and external facts, conflicts a believer would reduce as much as possible. The actuality of dissonance within a single human mind, however, may be significantly different. Dissonance seems to be an essential element in the functioning of the brain, not superimposed on a presumably consonant basic structure but so essential that one cannot imagine the removal of dissonance without the diminution or even extinction of mental function. One might say provocatively that cognition *is* dissonance, the factor intermodulating the centers within the brain and the external senses.

In his own way, William James expressed a version of this view in his use of the terms "*psychic overtone, suffusion,* or *fringe* to designate the influence of a faint brain-process upon our thought, as it makes it aware of relations and objects but dimly perceived."[38] James's "fringe of felt affinity and discord" relies "particularly [on] the relation of harmony and discord" to describe "a 'fringe' of unarticulated affinities" surrounding every feature of what he calls "the stream of consciousness."[39] His terminology explicitly brings forward dissonance ("felt affinity and discord") as a crucial descriptor of a brain process, whose intensity he diagrams (figure 14.5) and describes as being

Figure 14.5
William James's diagram of the intensity over time of a "brain-process," from his *Principles of Psychology*
(1890).

just like the "overtones" in music. Different instruments give the "same note," but each in a different
voice, because each gives more than that note, namely, various upper harmonics [overtones] of which
differ from one instrument to another. They are not separately heard by the ear; they blend with the
fundamental note, and suffuse it, and alter it; and even so do the waxing and waning brain-processes
at every moment blend with and suffuse and alter the psychic effect of the processes which are at
their culminating point.[40]

Thus, James's "fringe" really is a way of describing the constant intermixture of dissonances as they constitute what he calls the "stream of consciousness." Having studied
Helmholtz closely, James's diagram implicitly reflects the way a fringe of dissonance encircles every consonance. James's language of mingled cerebral "overtones" consciously
evokes an orchestra in which the instruments blend with each other in constantly shifting
configurations that allow the momentary emergence or receding of any one (or group) of
them as a principal voice. However different the various instruments, James insists on their
continual connection via the more or less dissonant "fringe" between them. In contrast,

the traditional psychology talks like one who should say a river consists of nothing but pailsful,
spoonsful, quartpotsful, barrelsful, and other molded forms of water. Even were the pails and pots
all actually standing in the stream, still between them the free water would continue to flow. It is just
this free water of consciousness that psychologists resolutely overlook. Every definite image in the
mind is steeped and dyed in the free water that flows round it. With it goes the sense of its relations,
near and remote, the dying echo of whence it came to us, the dawning sense of whither it is to lead.[41]

Earlier, we noted that a very large number of voices gives body to the sound of the whole,
as in Tallis's forty-part motet. In full cry, the hundred metronomes of Ligeti's *Poème symphonique* sound like a wild rainstorm. As the voices increase in number, one is less and less
able to follow any one of them; their overall effect becomes *texture*. In the search for ways
in which the polyphonic mind can achieve a kind of coherence or oneness, textural unity
may be important, especially the particular character of the texture.

Conversely, fundamental problems emerge from pathological states affecting the
polyphony of the brain. Cerebral trauma can sometimes disrupt its ability to integrate
polyphonic voices, as Oliver Sacks noted in his patient Rachael Y., a musician who, after severe head and spinal injuries, found herself in a condition he called *dysharmonia*.
Listening to Beethoven's String Quartet op. 131, she described hearing "four separate voices,

four thin, sharp laser beams, beaming in different directions. … And when I listen to an orchestra I hear twenty intense laser voices. It is extremely difficult to integrate all these different voices into some entity that makes sense."[42] Though she also lost her sense of perfect pitch, she experienced her dysharmonia as "a real torture," an even more fundamental and disorienting change in her inner world.

Closely related problems may be crucial to the condition commonly called attention-deficit/hyperactivity disorder (ADHD), characterized in the medical literature by various forms of inability to sustain attention, increased distractibility, hyperactivity, and impulsivity. Such difficulties with cognitive control also characterize major depression, schizophrenia, and Alzheimer's disease, not to speak of the multitasking "distracted minds" that seem to have become common.[43] When I asked a student of mine diagnosed with ADHD what he experienced, he told me that there were three different radios always going on in his mind, two of them talk shows, another playing music. If he tried to pay attention to a discussion happening in class, he struggled to disentangle it from these simultaneous inner "broadcasts" that distracted him. He felt that medication such as Ritalin "turned down" the radios' volumes sufficiently to permit attention to something other than those broadcasts. He also noticed that if the "music" radio played a song he knew, after he turned his attention to a conversation the song would take up just where he had left off. This characteristic distractibility appears in other accounts; one clinical study of fMRI scans concluded that indeed "attentional disregulation involves a large number of brain regions" including those "related to overall arousal and attention."[44] If so, we could understand ADHD as a hyper-awareness of the brain's inner polyphony that may also illuminate the underlying functioning of the brain during "normal" attention.

Let us return to the status of polyphony as an analogy. Analogy is not merely figurative language but has important uses in scientific practice, not just in teaching or popularization. As Gerald Holton pointed out, "modern science began with a quarrel over a metaphor," the metaphor of uniform circular motion that Copernicus upheld and Kepler changed.[45] Including under the general term "metaphor" all kinds of analogies and models, Holton describes the essential functions served by metaphor both among scientific practitioners and between them and the larger public. He instances Thomas Young's statement that the wave nature of light "is strongly confirmed by the analogy between the colours of a thin plate and the sounds of a series of organ pipes."[46] Likewise, Einstein considered that his "happiest thought" was an extended analogy between electromagnetic and gravitational fields.[47] For them and many other scientists, Holton concludes that "the urges to find analogies, and thereby to simplify and unify the various branches of a science, are actively at work in the background of [their] research."[48]

In the examples we have considered, polyphony has so often been used as an analogy that we may be justified in using the general concept of polyphonicity to describe the functioning of multiple parallel processes in the brain. We have seen some neuroscientists use terms like "orchestra" or "harmony" to illustrate their findings and make them more intuitively appealing. Beyond appeals to intuition, these comparisons rest on a thorough-going similarity between polyphony and the functioning of the brain, what mathematicians call an *isometric mapping*, a one-to-one correspondence between two different sets. I think these analogies to polyphony are so frequent because there are few, if any, other things in common experience to which the brain's operation can be fairly compared. For instance, despite some common hierarchical aspects, one can't analogize the brain to an army because it lacks a commander-in-chief; as we have seen, the homunculus and Cartesian Theater likewise are untenable analogies. Were other constructs to emerge that would be isometric to the brain's function, we should grant them special attention as giving us a new way to think about this almost unthinkably complex organ.

Recall von Neumann's argument that, in contrast to the linear tape spooling serially through a Turing machine, "large and efficient natural automata" like the brain "are likely to be highly *parallel*," to be polyphonic, in the terminology we have been considering.[49] In physics, wave and particle concepts seem so fundamental that we speak of a wave theory or particle theory of light or of matter, rather than a "wave analogy" or "particle analogy." If so, should we not speak of a polyphonic theory of the brain?

15 Music of the Hemispheres

In the twenty-first century, the study of the intertwined polyphonic rhythms of the brain remains active and vibrant. This final chapter contains a portrait of that ongoing project as presented by one of its principal exponents, the neuroscientist György Buzsáki. The study of brain rhythms confirms their fundamental significance for many aspects of brain function, such as sleep, arousal, and the overall constitution of the observable patterns of EEGs. In the view of many neuroscientists, these results have great promise for further development and arguably will remain of lasting significance for the understanding of the brain. Whatever new and unforeseen directions it may take in the future, neuroscience will need to address and illuminate the fundamental polyphony that now seems a central and inescapable aspect of human personhood and the brain that embodies it.

Buzsáki is an eminent researcher who received the Brain Prize in 2011, stands in the top 1 percent of the most-cited authors in neuroscience, and holds an endowed chair at New York University. His book *Rhythms of the Brain* (2006) gives an exceptionally lucid and detailed presentation of the contemporary neuroscience of brain rhythms.[1] Buzsáki cites over a thousand books and articles; he presents a broad picture that includes the work of many groups, as well as his own, careful to include difficulties as well as successes. As of 2016, his working hypothesis remains "that in brain networks, especially those serving cognitive functions, the packaging and segmentation of neural information is supported by the numerous self-organized rhythms the brain generates"; work in his lab "focuses largely on the generation of these various oscillations, their spatial and temporal relationships, and the role of inhibition in the enforcement of syntactic rules," meaning rules that take the "vocabulary" of the brain's constituent oscillations and from them show "the generation of virtually infinite combinations by finite numbers of lexical elements."[2]

These constituent neural oscillations come in many varieties, far more complex than anticipated by early hypotheses (such as Wiener's) of brain autocorrelation.[3] Beginning with Turing's 1936 program to explain the mind through pure computation, connectionists assumed that neural networks were composed of inherently passive constituents that would give rise to complex behavior only when linked together.[4] By the late 1980s, though, detailed research showed that neurons do far more than passively integrating their inputs.[5] In that light, Buzsáki is critical of such "top-down" approaches as artificial intelligence (AI) that disregard the "nuts and bolts" of the neural substrate or treat it as merely passive. He emphasizes that now "we consider the neuron to be a dynamic piece of machinery with enormous computational power … capable of generating a rich repertoire of activities, including oscillation and resonance at multiple frequencies."[6]

Like several other neuroscientists we have considered, Buzsáki often adverts to musical metaphors: "A neuron is a complicated resonator like a Stradivari violin, not necessarily tuned to a particular fundamental frequency, but endowed with a rich response repertory over a wide range of frequencies."[7] Indeed, a single neuron is more complex than a violin, whose characteristic spectrum of overtones is fairly constant over its range; "neurons, on the other hand, can dynamically change their resonant properties, as if the musician changed instrument between notes."[8] Even more, "every part of the neuron can function as a resonator-oscillator," so that "a single neuron consists of myriads of potential resonators."[9] Thus, in terms of our earlier discussion, each neuron deserves to be considered a "voice" in its own right within the larger polyphony of the brain; individual neurons also form parts of larger networks, which then can also function as important contributors to the whole polyphony.

These oscillations can even be observed in brain slices removed from the body. Thinly cut sections of the brain, perfused with oxygenated cerebrospinal fluid, can remain alive for several hours; in this way, the whole hippocampus of a rat can continue to function for a time in vitro. These slices show a variety of oscillatory patterns reminiscent of those in the intact brain even though only a small fraction of the network involved may have been included in the section.[10] Returning to the living brain, thin insulated electrodes can be inserted that can pick up the action potentials of a single neuron, if the conducting tip of the electrode is sufficiently close to that neuron. This long-standing experimental technique (introduced by David Hubel in 1957 to explore single neurons in the visual cortex) amplifies these signals through a loudspeaker, so that the neurons are literally *heard*, rather than seen. Buzsáki notes that, "with some maneuvering by a mechanical micromanipulator device, the signal can be maximized so that one neuron's 'voice' stands out from the others, a procedure called cell 'isolation.' If other active (spiking) neurons are in the vicinity of the tip, the electrode records all of them."[11] Thus, this procedure allows us to "hear" single neurons as well as a whole chorus of them, depending on the position of the electrode.

Because each neural "instrument" is so rich in itself, Buzsáki infers the concomitant complexity of the "orchestra" they compose when joined together. He notes that "inputs from the physical world" can act "much like the conductor's influence on the members of a symphony." This influence persists even "if the players are randomly placed in the concert hall, and if the musicians' ability to listen to others is attenuated by placing wax in their ears. … Conversely, the piece can also be played without the conductor, based exclusively on intraensemble interactions," such as those orchestras who play without a conductor.

"Similarly, the precise timing of central neurons depends not only on the physical inputs but also on the exchange of signals among central neuronal ensembles. These two sources of synchronization, externally driven and self-generated, should be distinguished because their tuning roles are often very different."[12] We will shortly consider examples of several kinds of "tuning."

The resulting neural symphony is far more than the mutual excitation of neurons and their circuits, which also interact through inhibition, especially via different kinds of "interneurons." As Buzsáki emphasizes, "in the absence of inhibition, any external input, weak or strong, would generate more or less the same one-way pattern, an avalanche of excitation involving the whole population" of neurons, which would resemble a kind of generalized seizure.[13] Instead, these systems of neurons often oscillate, mandated by the fundamental physics and mathematics that apply to all processes that balance excitation and inhibition.

To gauge the number of basic brain oscillators, Markku Penttonen and Buzsáki examined the rat's hippocampal rhythms, which comprise a continuous range of frequencies between 0.02 and 600 hertz—a far wider range than previously recognized—and included four slow bands not previously noted (figure 15.1). In light of these findings, they argued that "at least ten postulated distinct mechanisms are required to cover this large frequency range," over four orders of magnitude. Thus, there are at least ten oscillatory "voices" in the brain and, "most importantly, there is a definable relationship among all brain oscillators: a geometrical progression of mean frequencies from band to band with a roughly constant ratio of e, 2.17—the base of the natural (Napierian) logarithm." Because e is irrational, "the phase of coupled oscillators of the various bands will vary on each cycle forever, resulting in a nonrepeating, quasi-periodic or weakly chaotic pattern; this is the main characteristic of the EEG."[14] Much earlier, Oresme already thought that planetary orbits stand in irrational ratios so that celestial configurations likewise would never repeat, which he and Kepler took as evidence of the endless novelty and life in the cosmos. The irrational tempo ratios of Nancarrow's Study no. 33 also involve this unrepeatability.

Buzsáki also notes that these oscillators require time coordination between them in order to exert their impact. Further, psychological phenomena (such as successively recognizing someone's face, then recalling his or her name, profession, and common friends) seem to require hierarchical processing, as well as "the engagement of neuronal networks at multiple spatial scales. ... Because many of these oscillators are active simultaneously, we can conclude that the brain operates at multiple time scales."[15] Then too, "the brain does not operate continuously but discontiguously, using temporal packages or quanta. ... The wave length of the oscillatory category determines the temporal windows of processing," which Buzsáki connects to William James's concept of the segmentation of experience. James considered the unit of experience to be "a duration, with a bow and a stern, as it were—a rearward and a forward-looking end. It is only as parts of the duration-block that the relation of succession of one end to the other is perceived. ... We seem to feel the interval

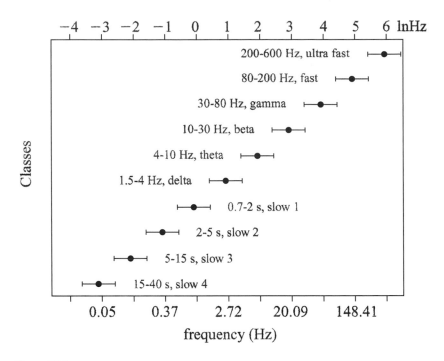

Figure 15.1
Oscillatory classes of brain rhythms as observed in mice by Penttonen and Buzsáki (2003), who concluded that each band would require a separate oscillatory structure, thus at least ten oscillators in all.

of time as a whole, with its two ends embedded in it."[16] Buzsáki deduces from this temporal packaging that "different frequencies favor different kinds of connections and different levels of computation. In general, slow oscillators can involve many neurons in large brain areas, whereas the short time windows of fast oscillators facilitate local integration, largely because of the limitations of the axon conduction delays."[17]

The unperturbed brain, resting or sleeping, invites us to contemplate the connected operation of these various oscillators in generating an overall cerebral state from within, in the absence of external stimuli. This "default state" of the brain is not constant but comprises five stages of sleep, each with its own oscillatory rhythms. Thus, sleep is a composition of many rhythms arranged in a larger overall rhythm, comprising a larger amount of slow wave sleep (SWS) and a smaller fraction of rapid eye movement (REM) sleep (figure 15.2). Though the real physiological purpose of sleep remains unclear, Buzsáki's "two-stage model of memory trace formation" is one of his most cited contributions, combining ways in which sleep restores the homeostasis of the brain as well as consolidating memories: Brendon O. Watson and he argue that "during sleep—that is, after waking

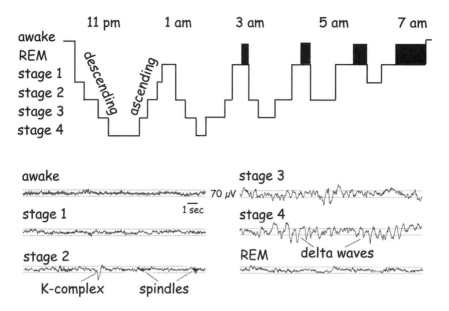

Figure 15.2
Buzsáki (2006, 188) describes sleep: "a dampened oscillation with approximately 90-minute periods. Top: Hypnogram of a night's sleep in a young adult human. Note periodic ascending and descending phases. Bottom: Representative scalp-recorded EEG segments from each sleep stage."

acquisition—memories are not wiped away or simply made to decay less slowly, but are often actually improved, molded, and shaped."[18]

In this process, they argue that slow wave and REM sleep both play important roles, especially via one particular oscillation,

the sharp-wave ripple (SPW-R): a brief (50–150 millisecond) electrical rhythm generated by an intrinsically self-organizing process in the hippocampus, which apparently provides a perfect mechanism for the precise consolidation of waking experience. This brief rhythm is cross-frequency-coupled with other rhythms such as slow waves and sleep spindles, and it represents the most synchronous physiological pattern in the mammalian brain: 10 to 18 percent of all neurons in the hippocampus and highly interconnected regions (subiculum and entorhinal cortex) discharge during these events.[19]

Indeed, "artificially induced SPW-R-like patterns can strengthen synapses even in brain slices *in vitro*," further indicating the significance of these patterns in memory formation. Albert K. Lee and Matthew A. Wilson (2002) observed that "sequences of neuronal assemblies present during waking behavior are replayed (and at higher speed) during subsequent SPW-R events," indicating that these ripples correspond to a rat who had run along a track to a food well and then replayed those memories first when awake and then again during slow wave sleep (figure 15.3).[20] Further, Gabrielle Girardeau and her collaborators (2009)

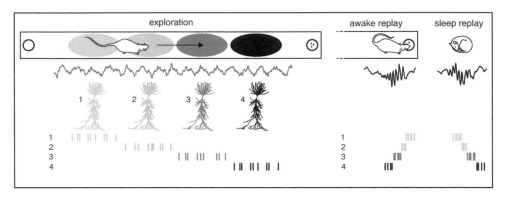

Figure 15.3
Sharp wave-ripple (SPW-R) neuronal replay (from Girardeau and Zugaro 2011). When a rat is running on a linear track, the hippocampus oscillates at theta frequency (shown below its route) and place cells are successively activated as the rat enters their respective place fields (ellipses), yielding a neuronal sequence (vertical ticks). Arriving at the food well, the rat stops to consume a reward and the place cell sequence reactivates in *reverse* order (reverse replay) during SPW-Rs. In subsequent slow wave sleep and quiet rest periods, place cells reactivate in the same order as during exploration (forward replay).

were able to suppress these ripples during the sleep of one group of rats, whose memory formation was significantly lower than that of a control group allowed to sleep experiencing the ripples that enabled them to replay and hence reinforce their experience.[21]

In these experiments, one seems to observe (and even change) the inner life of the rat directly via its brain rhythms. Watson and Buzsáki note that, "while SPW-Rs can emerge from a self-organizing process, they are often biased to occur by other brain-wide oscillations, including sleep states" at various rhythms, including thalamo-cortical sleep-spindle oscillations (12–16 hertz), slow oscillations, as well as ultraslow waves (0.1 hertz), all modulating each other.[22] Thus, one sees in this example the significant interaction of all these rhythms to create the whole polyphony. Even more striking, several 2011 experiments showed that "groups of neurons that fire in a specific sequence during a novel running behavior will have actually fired in that same order (though compressed in time) in sleep *before* the new behavior occurs. This suggests that the brain may use pre-constructed network structures during behavior, and that sleep plays a role in the pre-behavior activity of those network constructs."[23] As musical polyphony can draw on a preconstructed motive or theme, even what seems new in our experience may rely on structures long prepared in our brains.

Again using musical metaphors, Watson and Buzsáki sum up their theory "that sleep is a general tool used by mammals to tune their entire neural system to be able to properly acquire, select, and store information."[24] More generally, they emphasize that, as all mammalian brains share the same types of neurons, they also share the same basic neural rhythms, even though over the course of mammalian evolution, brain size has increased several thousandfold.[25] "Every known pattern of local field potential, oscillatory or intermittent, in the human brain is present in other mammals investigated to date [2013]. Not only the frequency bands but also the temporal aspects of oscillatory activity (such as duration and temporal evolution) and, importantly, their behavioral correlations are conserved."[26] Brain rhythms are robust and heritable phenotypes, thus strongly genetically controlled; "the existing knowledge clearly reveals that brain rhythms are among the most heritable traits in mammals, leading to the suggestion that EEG patterns could be used for 'fingerprinting' individuals."[27] Then too, a number of neurological and mental diseases can be characterized through the use of EEGs and magnetoencephalograms (MEGs), most obviously in the various forms of epilepsy, chorea, or Huntington and Parkinson's diseases in which symptoms such as palsy or tremor seem closely correlated with brain rhythms.[28] Besides these, there is considerable evidence connecting other mental conditions such as schizophrenia with brain dysrhythmias and "oscillopathies."[29]

As we move from the sleeping brain into wakefulness, the symphony of rhythms shifts markedly, already evident in a general way in the transition from alpha to beta waves in the EEG. For example, consider the *mu rhythm*, named after its characteristic mu-shaped (μ) waveform in the EEG. Similar to alpha in frequency, mu responds differently: though alpha is interrupted when the subject's eyes are opened (or even if the subject begins to make mental arithmetic calculations), mu is indifferent to the presence or absence of visual inputs but requires immobility of the skeletal muscles. The mu rhythm can be blocked merely by moving a finger or toe. Though more general forms of the alpha rhythm occurs at many places in the brain, mu oscillations seem to have a localized source on the banks of the Rolandic fissure, the major groove (sulcus) that separates the frontal and parietal lobes of the brain; you can locate this fissure roughly by tracing a line directly over the top of your head connecting your ears. The disappearance of the mu rhythm seems directly correlated not only with motion but even with the *intention* to move. As Buzsáki notes, "even in the case of a very simple movement, such as moving a finger voluntarily, attenuation of the mu rhythm begins approximately 2 seconds prior to the actual movement, indicating time-intense computation of a voluntary act."[30] This surprisingly long time preceding actual movement also indicates how long an intention is being rhythmically prepared in the brain even before one becomes aware of that intention: it does not feel like it takes two seconds to move a finger. Indeed, experiments by Gert Pfurtscheller and Andrea Berghold (1989) showed that local alpha and mu rhythms could be topographically dissociated even when the subject "planned" to move, *but did not move*.[31] Here we confront evidence that a change in brain rhythm arguably represents and precedes action, representing the act of thought itself, whether or not the action contemplated is carried out.

Because of these observable connections even with the mere intention to move, mu rhythms have been used in the development of brain-computer interfaces (BCI), which allow direct linkage between a subject's EEG and a computer. This is of great importance for those afflicted by diseases such as amyotrophic lateral sclerosis (ALS, or Lou Gehrig's disease) who are almost totally paralyzed; these individuals can now be trained to use their brain rhythms to communicate via electronic keyboards or speech devices. In 2013, Karl La Fleur and his colleagues showed that able-bodied participants could control a small quadcopter in three-dimensional physical space (an indoor basketball court) solely via an EEG attached to their scalp, connected to a computer that sent wireless signals to the quadcopter. The participants had to pass through several stages of learning to control the motion of a dot on a screen (first in one, then in two dimensions) using the EEG until finally they were able to use mental imagery of moving their hands to control the quadcopter, which flew outside their field of view; they only saw the image provided by an on-board camera, as if they sat in the pilot's seat. After training, they were able to pilot the quad-copter through a number of rings, having to decide on a flight path without constraints or prompts.[32] To do so, the users had to learn to redirect their mental polyphony in new ways, learning to connect inner states with the external motion of the quadcopter but without the intermediacy of the body. The difficulty of this process indicates how profoundly the body mediates the brain's activity.

In contrast, newborns emerging from their mostly sleeplike prenatal state use the exercise of their bodily motions to connect their inner states with the external world. As Buzsáki points out, "there is no a priori reason why representation of the environment in the brain should be three-dimensional and linear [in time], instead of n-dimensional and logarithmic or exponential. The brain's sensory representations acquire real-world metrics early in development," through the motor-activity-generated sensations that begin late in intrauterine life.[33] This is accomplished through interconnecting various kinds of oscillations that connect motor experience with neural states; indeed, "the first organized pattern in the intact neocortex is a rhythm."[34]

In these processes, the various oscillations interact in several ways so as to create temporal synchrony among various (sometimes widely spaced) populations of neurons. Here, and at many other situations in the brain, that synchrony seems to provide a practical way of binding together relevant information from many neurons: for example, "neurons that fire together in response to external visual patterns will wire together to form functional assemblies."[35] These interactions among oscillators follow basic patterns first described in physics, such as mutual entrainment (as when a child swinging excites a nearby empty swing to begin moving) and phase-locking (as when two children come to swing with a fixed delay between their relative positions within each of their cycles). In the nervous system, "phase reset" has particular importance: a sudden impulse can cause oscillators to "reset" to the beginning of their cycles. Such impulses can be provided by the kinds of behavioral stimulus that act on the developing organism as it experiences the world.[36] Further, the synchrony provided by oscillatory processes does not cost the organism much energy. The sheer multiplicity of these processes and their interactions shapes the brain at many levels; for instance, slower rhythms (involving larger areas of the brain) will often modulate the behavior of faster rhythms that are more local.

In the "awakened" brain, gamma oscillations seem to have particular importance; characteristically, these oscillations are localized temporarily in cerebral areas engaged in a particular operation, rather than appearing throughout the brain. For instance, a learned association between a visual and tactile stimulus will evoke marked gamma oscillations coherently over the visual and sensory cortexes between 150 and 300 milliseconds after the onset of the visual stimulus, seemingly at the point when that stimulus acquires meaning.[37] With this in mind, we return to Wolf Singer's theory (discussed in the previous chapter) that gamma oscillations might provide the solution to the fundamental binding problem, how different features of an experience (such as a visual image and a tactile sensation, associated with different groups of neurons widely separated in the brain) can be bound together so quickly (within about 200 milliseconds). Here different modes of explanation remain under consideration. Some researchers interpret hierarchical aspects of brain structure (such as the various levels of visual processing) as implying "gnostic" neurons that contain the highest-level information forwarded up the hierarchy of processing each experience or object.[38] In Buzsáki's view, the main difficulty with this argument is that

there is no top in the brain hierarchy and the bottom-up paths are always coupled to top-down connections. So the gnostic neurons would inevitably end up sending their impulses back to earlier processing stages. But what is the point of sending impulses back to neurons representing the elements of an object after the whole has been already identified by the gnostic elements? Because there are no stop signals in the bottom-up model, it is not clear what mechanisms would prevent the reverberation of activity in an interconnected system after the object is recognized.[39]

Further, "it remains to be explained how the action systems of the brain can be mobilized effectively by a handful of decision-making neurons."

Such basic difficulties seem to counterindicate the efficacy of gnostic neurons and leave only some kind of collective binding through various oscillatory processes, though these two explanatory strategies—the gnostic and the collective—may not be mutually exclusive. To be sure, problems emerged with the simplest versions of Singer's suggestion that gamma oscillations might be the neural correlate of visual awareness. Though he and Francis Crick had been early proponents of this idea, by 2016 Christof Koch and his collaborators expressed disappointment with that theory as "not showing any predictive power" because "gamma synchrony can occur in the absence of consciousness" and hence "gamma band oscillations are not necessary for seeing" and are "illusory ... as signatures of consciousness."[40] Instead, they concluded that, though "no single brain area seems necessary for consciousness ... a few areas, especially in the posterior cortical hot zone are good candidates for both full and content-specific NCC [neural correlates of consciousness]."[41] In response, Buzsáki notes that, though gamma oscillations may be present without consciousness in certain situations (such as in the olfactory bulb or during the "up" phase of slow oscillations), "there is no consciousness in the absence of gamma oscillations at the right place in the brain because gamma is only a reflection of the rapid dialogue between excitatory and inhibitory interneurons. If the dialogue is on, gamma is on. If the dialogue is not on, there is no local processing of information, thus no function, no consciousness."[42]

For such a complex system as the brain, the simplest versions of such an overarching theory face many difficulties and will doubtless require much further work. Nevertheless, many provocative findings have emerged about the larger significance of gamma oscillations that will probably remain of great importance. As Buzsáki points out, "even if gamma oscillation proves to be irrelevant for the binding problem, the oscillation remains a central timing mechanism essential for synaptic plasticity," the ability of neurons to change their connections, sometimes through the motion of dendritic spines, sometimes by replacing neurons. If so, "gamma oscillations may link the problem of binding to plasticity. … The assembly bound together by gamma-oscillation-induced synchrony can reconstruct patterns on the basis of partial cues because of the temporally fortified connections among neuron assembly members."[43]

Buzsáki's own approach seeks not just the *vocabulary* of brain activity but its *syntax*, the ways in which neural "words" form meaningful "sentences."[44] Here musical and linguistic approaches may illuminate each other; Buzsáki speculates that "the roots of language and musical syntax do in fact emanate from the neural syntax native to the brain, since it is the neural syntax that secures a match between brains, which both generate and interpret information."[45] Polyphony offers concepts that complement those drawn from the one-dimensional stream of verbal syntax, which is more oriented to sequential words rather than simultaneous processes. For instance, Jean-Pierre Changeux and his collaborators model the parallel processing of signals as a "neuronal workspace" in which the "workspace neurons generate time-ordered sequences, or 'melodies,' of mental representations." Each "note" of these melodies "is held on-line in working memory until a moment of convergence or resolution is reached," which Changeux describes as a "chord."[46]

The emergence of these "neuronal melodies" from the underlying cerebral counterpoint illuminates a central problem von Neumann raised, which we looked at in the previous chapter: how could the brain's parallel processing eventuate in an apparently linear stream of thought, the tape unrolling from a Turing machine? As we have repeatedly noticed, complex counterpoint can still produce a simple "surface" or melody. In his own way, Cage noticed this even in the context of his radically open and indeterminate compositions, whose silences at first "seemed very long and the sounds seemed really separated in space." But after a while (he recounted), "no matter what we do, it ends by being melodic."[47]

Similar conceptual approaches may also describe a complex hierarchy in which some neurons have a centralizing function yet remain interdependent with other neurons. Confronting such processes in neuronal assemblies, Amiram Grinvald and collaborators (2003) described the single cortical neurons as "obedient members of a huge orchestra" that together produce "the sum of the individual notes, the tunes played by all instruments"; they describe their problem as "trying to understand what the orchestra is playing when we do not have a partitura [score] in front of us."[48] They go on to describe their experiments as first studying "the role the neuron plays in the orchestra, when we let the orchestra play a familiar tune by presenting it with a stimulus. We could then look at whether the neuron played the same role in the orchestra, once the tune was no longer imposed, i.e. in the absence of stimulation."[49] They conclude that the single neurons are not "soloists" but "well-tuned obedient players." Their paper illustrates yet again that this kind of orchestral terminology appears in neuroscience papers, not just popularized accounts.[50]

The precise details of polyphonic performance may also be significant for neuroscientific discourse, including the ways in which musical language tends to be used or avoided. Consider the operation of the stomatogastric ganglion of the lobster, studied to illustrate the interrelation of neurons in invertebrates. The original 1983 research papers by P. S. Dickinson and F. Nagy described their findings purely in physiological terms, apart from occasional references to the rhythms of the systems they were studying.[51] But a 2010 technical review article by Yves Frégnac and his associates described this work as showing that certain cells

play the role of "orchestra leaders" whose activity triggers the widespread diffusion of neuromodulators. This neuromodulation impacts on the intrinsic reactivity of the other cells by changing reversibly the expressed repertoire of membrane conductances. Consequently, the individual excitability patterns of any given cell will change depending on the context (before, during, or after the orchestra leader has fired). ... These data, often ignored from the vertebrate research community, support the hypothesis of "orchestra leader"-like neurons which, through the presence or absence of firing activity, condition and format merging and representation across functional assemblies.[52]

This account depends on the awareness that the leader of an instrumental section (the principal violist, for example, the "first chair" or "section leader," in American parlance) will coordinate the playing of the whole section (its attacks, bowing, phrasing) in rehearsal and performance.[53] Such a "leader" plays along with the section in such a way as both to lead as well as to participate in their collective playing.[54] In fact, this usage depends rather sensitively on the differences between orchestral playing and the polyphony of solo voices, each playing a single part, as in chamber music. Indeed, the problems of the brain (or even the nervous system of a lobster) are more like orchestral playing than chamber music; the mass of second violins need a sublevel of coordination provided by the section leader, as apparently does the lobster ganglion and even more so the vertebrate brain.

This example illustrates the way polyphonic terminology seem increasingly appropriate and useful as the technical biological findings take the form of a more general theory. In that more interpretative discourse, notions like "orchestra leader" are not merely popularizing (for it appeared in a technical overview directed to neuroscientists) but rather offer ways to clarify concepts of organization and coordination that prove easier to express in music than via other analogies. After all, the orchestral playing of a viola section involves subtle forms of mutual coordination not really equivalent to common social units like "teams" or "departments" or "families." Then too, once the concept of "rhythm" has been taken as fundamental (even in technical papers), it seems rather natural to use other concepts likewise drawn from music.

In the search for ways to coordinate neural systems without a single "conductor," it should also be borne in mind that contemporary conductorless orchestras (like their predecessors in earlier centuries) can function well through sensitive signaling, leading, and following among the section leaders, who in turn coordinate their sections to participate musically in the whole. To be sure, these ensembles depend on musicians who are not slavishly dependent on an external beat but are capable of signaling and responding flexibly; likewise, the complexity of each single neuron—its multiple resonances and excitations—seems to be an essential ingredient for the conductorless coordination of the whole brain.

These concepts offer a middle way between the now-abandoned search for the cerebral "conductor" and the total democracy (if not idealized anarchy) of pure connectionism. Many contemporary neuroscientists seem to find both extremes untenable, though the dream of radical neuronal equality faded more recently. Yet elements of both these extremes will probably inform more mature versions. From the ancient dream of a central intelligence ruling somewhere within the brain, important aspects of neuronal hierarchy remain, such as the "orchestra leader" neurons. From the radical vision of neuronal "democracy" remains the overarching interconnection of the brain, the absence of any "top" level not connected to a "bottom."

As of this writing, the neuroscience of polyphony has been relatively little explored and may repay further examination; compared to many studies of the brain's response to music in general, rather few have examined the problems of polyphony in particular.[55] One of these noted that listening to polyphonic music recruited large-scale involvement of the brain, including "neural circuits underlying multiple forms of working memory, attention, semantic processing, target detection, and motor imagery. Thus, attentive listening to music appears to be enabled by areas that serve general functions, rather than by music-specific cortical modules."[56] Yet how does this general activity of the brain deal with the problem of integrating many voices in the listener's awareness? What enables the brain to process many-voiced music in ways that it cannot assimilate multiple conversations? Further investigation may illuminate not only the paradoxes of polyphony but the polyphonic brain itself.

Then too, neuroscientific treatments so far have generally relied on rather traditional aspects of polyphony, such as "orchestra leaders" or "neuronal melodies," as if only those aspects were familiar to the neuroscientists as they sought a new vocabulary to describe the "neural orchestra." As useful as those traditional references may be, one wonders whether a broader and more comprehensive acquaintance might open new doors in understanding the brain in terms of the language of polyphony. Especially given the enormous complexity of the brain, it seems likely that terms drawn from the polyphony of many voices (rather than the three or four voices usual in common-practice choral or orchestral music) may be of particular value. Here one thinks of Ligeti's micropolyphony of a hundred voices or the textural effects in the forty-part works of Tallis and Striggio. Could consciousness relate to its neural circuits as does the continuo to Striggio's forty voices above it (figures 7.1, 7.2)?

We have focused on the notated tradition of Western polyphony, mainly because its written transmission (and its related theoretical literature) has allowed us to discuss it more easily. But the "neural orchestra" does not play from a score, and thus other traditions of improvised polyphony may be even more relevant. In the Western tradition, the improvisational aspects of jazz make it especially appropriate. Still, much of standard jazz consists of improvised variations on a certain song or chord sequence (such as the blues), so perhaps we should rather consider free jazz as it developed since the 1950s, unfolding more improvisationally without prewritten chord changes.[57] Indeed, some have described free jazz in terms of the mathematical technique of swarm analysis, which has also been applied to self-organizing artificial intelligence.[58] Then too, once we consider improvised polyphony, we should return to such complex and long-developed practices as the polyphonic singing and playing of the Central African Pygmies, such as the Banda-Linda orchestra mentioned in chapter 1. By virtue of its large number of parts (eighteen in figure 1.3), this Pygmy music may offer comparisons that could allow us to articulate and understand more clearly the neural orchestra.[59] In any case, the insights derived from over a millennium of Western polyphony provide an important point of reference for describing all kinds of polyphonic practice.

At this point, we must leave this ongoing and necessarily incomplete story, which is still very much in the process of unfolding and may only become clear centuries hence. Whatever might be the more mature shape of that understanding, in twenty-first-century neuroscience—at least as practiced by Buzsáki and other like-minded researchers—brain rhythms remain a fruitful and important topic almost a century after the first EEGs. For the foreseeable future, neuroscientists will be engaged in studying how multiple neural subrhythms sum to form the overall rhythm manifest as EEG traces and (ultimately) as various mental states, the polyphonic "music of the hemispheres" that now seems inextricable from human personhood.

Postlude

In April 2016, I witnessed an outdoor performance of John Luther Adams's *Sila: The Breath of the World* (2014). On a beautiful spring day in North Carolina, about a hundred young musicians were spread widely across a large grassy area between university buildings. Five choirs of voices and instruments were loosely grouped in different areas; the players stood or sat somewhat separated from each other, allowing plenty of space for listeners to walk among them or find a place to stand or sit. Each player had a separate part, "traversing sixteen harmonic clouds, grounded on the first sixteen harmonic overtones of a low B-flat," as the composer specifies.[1] Because he wished these chords to be in just intonation, "off the grid of twelve-tone equal temperament," performers' smartphones helped them find the "natural" overtones provided through their earbuds. The idyllic outdoor setting delicately balanced these electronic guardians of the "natural." Adams explains that, "like the tuning, the flow of musical time in *Sila* is also off the grid. There is no conductor. Each musician is a soloist who plays or sings a unique part at her or his own pace. The sequence of musical events is composed, but the length of each event is flexible. The music breathes. ... *Sila* comes out of the earth and rises up to the sky."

The pace of events was slow and dreamy, the sounds generally quiet, though sometimes swelling in volume; the sixteen "harmonic clouds" took about an hour to unfold very gradually. Each player chose when to play the next note in the chord without any signal or coordination from the others. They did so with notable calm and deliberation, often with long pauses between notes. It seemed they took care finding the "natural" overtone they needed to play, which often would involve having to tune "in the cracks" between the equal half-steps built into many instruments. In the process, they listened to the whole ensemble as well as to the whispered help of their smartphones. Each performer's individual responsibility seemed to evoke a special kind of dignity somewhat different from the usual deportment of musicians under a conductor's direction. One could see and feel this as one wandered among the players, able to observe them closely as each separately waited, listened, gathered themselves, and then finally sounded or sang her or his next note.

It was fascinating to observe the individuality of each player, how they carried themselves, who they were, how they decided to shape their sounds. Standing among them, I was reminded of Janet Cardiff's sound installation of the Tallis forty-part motet, yet now one approached not loudspeakers but real persons who also (one sensed) were responding to the listeners among them as well as to the sound of the whole, including the ambient sounds of that spring day (♪ sound example 16.1). At some places in the field, one could hear the whole ensemble in different ways, with its various instruments and groups in continuously shifting balance.

The event had an extraordinary aura. Listeners as well as players seemed caught up in something profound and tranquil. Wondering at this, a friend asked: "Is this the end of the world?" Indeed, one was not sure when the performance had begun or when it would end, for (the composer writes) "there is no demarcated ending. As the music of the performance gradually dissolves into the larger sonic landscape, the musicians join the audience in listening to the continuing music of the place."

In its own way, *Sila* shows the continuing presence of polyphony in the twenty-first century. Not seeking expression, it uses chords based on overtones and just intonation to evoke the cosmic calm of ancient musical science. Its multiphony follows the unstructured example of Cage's music, imposing no rhythmic unity, choosing polyphony without traditional counterpoint. Earlier in this book, the case of Nancarrow's Study no. 33—Canon $\sqrt{2}\,/\,2$ suggested that rhythmic irrationality would lead to a breakdown of the listener's ability to hold its two voices together, creating an impression of restless energy. In contrast, *Sila*, lacking any imposed rhythmic coordination, was tranquil and easily apprehended. In part, this was a result of its gentle harmonies, the absence of overt confrontation or dissonance, which reflected the composer's directions as well as the players' assent. Something about the beauty of that day also seemed to move and shape the experience, especially because the listener could move around in the field, hearing the individuals and the ensemble from different points of view. No recording could do this justice. Webern's notion of musical "space" and the additional dimensions of polyphony take on new meaning when the listener's body can be mobilized.

Spoken polyphony continues to take on new life in unexpected contexts. In a scene of the film *Manchester by the Sea* (2016), two characters encounter each other after years of alienation and separation. Stricken by tragedy and remorse, they address each other at first hesitantly and haltingly, with awkward silences, then so carried away by emotion that neither can wait their turn to speak. Their voices overlay uncontrollably, each pouring out long pent-up feelings while simultaneously reacting to the other, asking forgiveness and expressing anguish. Far from seeming cacophonous or jumbled, this spoken polyphony has an effect different from but no less moving than two sung voices. Indeed, the broken quality of their simultaneous speaking achieves a new kind of expressive force.

Having so much addressed the mind and brain, in closing I would like to reemphasize the fundamental significance of the body for polyphony.[2] At many points in this book, the body underlies what might seem to be the most cerebral and inward mental experiences.

The experience of dissonance is first of all a felt vibration in the delicate cilia of the inner ear, its resonance in the most physical sense. Bach's keyboard polyphony requires hands—and thumbs—not just theoretical contrapuntal science. Cage's story about hearing high- and low-pitched sounds—his nervous and circulatory systems—in the "soundless" anechoic chamber reminds us that the polyphony of the body, the continuo of life itself, remains even when all other sounds are eliminated. Through its multiple abilities to move in different ways at once, the body is inherently polymodal, even more than the physically motionless brain. Of course, body and nervous system are bound together inextricably; the intrinsic "many-voicedness" of the body and its mind shape each other in ways formed by eons of evolution.

The coordination of mind and body helps us return to our opening question about polyphony: how can we hear many voices at the same time? Oresme argued that we can hear several voices as having a larger unity if their pitches and rhythms (as well as each voice's microscopic structure of sound and silence) have sufficiently simple numerical relationships the mind can readily grasp, such as 1:2, 2:3, 3:4, and so on. Yet these "mental" concepts come from the interaction between body and brain. Even though 99.9 percent of the brain's neurons are connected only to its own internal functions, it strains credibility that the brain could form a concept such as "2:1" without prior sensory experience through the remaining 0.1 percent of neurons connected to the body. Not only does the brain need bodily experience of "twoness" in the static sense (such as touching two objects) but also of moving twice as fast as well as of hearing two pitches an octave (2:1) apart. Though purely visual stimuli are surely important, they can never convey the primordial experience of rhythm so essential to the brain, whose activities seem deeply connected with various interacting rhythms. In the womb, before any visual experience, we feel the rhythm of our mother's heart.

This intervention of the body illuminates the problem I raised at the beginning about how a pianist can play several lines of music at once. We can only do so because each hand is capable of learning to shape a single melodic line in time so that then both hands can learn to move together in rhythm, each playing its own line. The mind cannot learn polyphony without the body. To test this, try learning to play an unfamiliar polyrhythm purely mentally, without moving a muscle; not even a skilled musician can accomplish this, not even someone able to memorize a score without playing it. As we saw earlier, Chopin's Nouvelle étude no. 1 (figure 11.4) demonstrates how the hands teach the mind as the mind monitors the hands.

In this process, rhythm is primary, far more than pitch or other elements of music, reflecting the primacy of rhythm for both body and brain. Thus, one should understand "music" to comprise dance as well, because rhythm cannot be apprehended without dance, without the body in motion. These bodily issues of rhythm go deep into what we mean by time. In 1968, while investigating psychoacoustics at the electronic studio of the West German Radio, the composer Gottfried Michael Koenig discovered that if successive pitches are made to sound at a rate faster than twenty times per second, they are no longer heard as melody but coalesce into chords.[3] Koenig created this effect through artificially manipulated sound; in 2013, the young Canadian drummer Tom Grosset played an average of twenty beats per second for an entire minute, setting a new record.[4] Listening to his feat, one indeed seems to hear simultaneous rather than sequential sounds, as if he were sustaining a kind of "chord" of the two notes of his drum (♪ sound example 16.2).

Thus, the distinction between successive and simultaneous, so often taken as fundamental, becomes blurred at high tempi, above metronome markings of 1,200 beats per minute (most metronomes top out at a "prestissimo" of 208). This effect is even clearer in electric guitar renditions of Nikolai Rimsky-Korsakov's "Flight of the Bumblebee" that have reached a dizzying 1,300 beats per minute (as of 2012), at which tempo the song lasts only a few seconds, any sense of its "melody" mostly lost in a wash of harmony.[5] Still, the speeds involved are proportionately very slow in comparison to now-familiar relativistic manifestations of the mutual dependence of space, time, and motion, whose characteristic scale is the velocity of light and which therefore are quite different in cause and magnitude.

Here we stand at the origins of time, rhythm, and harmony but now in the realm of human physiology, rather than physics.[6] Beyond instrumental tours de force at extreme speed, the larger implication is that our experiences of simultaneity and succession are constructed by the brain at the smallest scale of neural functioning. Some experimental evidence indicates that the brain constructs its own sense of time, on the millisecond level, stitching together memories (and ongoing experience) empirically rather than relying on any hypothetical built-in pacemaker or timing circuits.[7] If so, the interplay of brain rhythms structures our sense of time so that the polyphonic "neural orchestra" apparently produces the seemingly linear unfolding of our experience, the "neural melodies." Through polyphony, body and mind become person. What began as a quest to imagine the mind of God has become the exploration of our own polyphonic persons.

Notes

Prelude

1. Pesic 2014b.

2. See Cherry 1953 for the initial experiments regarding this problem.

3. Gazzaley and Rosen 2016, 108, which gives a detailed account of the neuroscience of distraction.

4. Ibid., 130–132; for the problems of multitasking in aviation, see Loukoupoulos, Dismukes, and Barshi 2009; Jackson 2008 offers a more general perspective on distraction.

5. *Topics*, II.10, 114b34 (Aristotle 1984, 1:191); see also his discussion of simultaneous sense-perception (including hearing) in *Sense and Sensibilia* 447a11–449b4 (Aristotle 1984, 1:709–713), discussed below (43). Unless otherwise indicated (as in this case), all translations are by the author.

6. Plato, *Republic*, 580d–e (Plato 1997, 1188).

7. For instance, Burney (1957, 1:469) in 1782 mentions "polyphonic music, or counterpoint," which is among the earliest uses of "polyphony" in English recorded in the *Oxford English Dictionary*. For the thirteenth-century source, see Page 2007, 30–32, 124–128.

8. Cited in Arom 1991, 71.

9. For instance, Lomax (1968, 165) defines polyphony in his quantitative cross-cultural "cantometric" studies as "the use of simultaneously produced intervals other than in unison." He provides a typology of polyphonic types ranging from "no polyphony," to "drone polyphony," "isolated chords," "parallel chords," "harmony," and finally "counterpoint, two or more parts which are rhythmically and melodically independent" (65). Arom (1991, 37–93) outlines the definitions of polyphony and polyrhythm in Central African music.

10. Galilei 2003, 223–224.

11. Pesic 2014b, 9–20.

12. Cf. the different (but also significant) distinction between "gnostic" and "drastic" music made in Jankélévitch 2003, 77, and elaborated in Abbate 2004.

13. Mathiesen 1999; Haar 1961; Sullivan 1997; Wright 1989; Ciabattoni 2010; Macey 1998; Wegman 2005; Harrington 1987.

1 Global Contexts and Ancient Origins

1. In 1885, Alexander Ellis noted that in Javanese music "many instruments play together, but there is no harmony" (Helmholtz 1954, 237n3). Around 1890, Jan P. N. Land and Isaac Groneman studied Javanese gamelan music, which is polyphonic; see Terwen and Staples 2005. For the role of Javanese music in the development of Western music theory, see Perlman 2004. Among the founders of ethnomusicology, Adler 1886 only treats the history of polyphony in Western sources, but Adler 1908 does discuss Javanese music and its "polyphony without rules." For the perspective from 1911, see Stumpf 2012, 101–106.

2. The earliest Western history of world polyphony was Schneider 1969 (first published in 1934). For a helpful overview of polyphony in ethnomusicological context, see Sachs 1962, 175–191, who discusses and critiques

Schneider's work on 96, 127, 334–335. For the history of ethnomusicology, see Nettl 2010, treating Schneider on 115, 179. I thank Bruno Nettl for helpful correspondence on this issue.

3. For an example in Oceania, see Crowe 1981.

4. See Arom 1991 for a detailed description of central African Pygmy music. Chernoff 1981 gives a sensitive analysis of African rhythmic practices within a larger view of African society, discussing polyrhythm particularly on 44–60. See also Agawu 1995. For a cross-cultural perspective on polyrhythm, see Sachs 1962, 192–199.

5. For detailed discussion, see Arom 1991, 309–310.

6. Chernoff 1981, 51.

7. Lewis 2013, 53.

8. Ibid., 63.

9. Turnbull, as quoted in Lomax 1962, 437.

10. Lomax 1962, 1968.

11. Jordania 2006, 2015, as discussed by Nettl 2015, 46.

12. See Kilmer 1974; the song was published as "A Hurrian Cult Song from Ancient Ugarit" in Kilmer, Brown, and Crocker 1976. For a critique of this interpretation, see West 1994, to which Crocker 2011 responds.

13. Lomax 1968, 165–166, defining counterpoint on 65.

14. Ibid., 167–168.

15. Ibid., 167. Regarding cantometrics, Lomax's system for classifying aspects of musical practice and social organization, see Nettl 2015, 96–97, 119–120; for alternatives to Lomax's views about the social correlates of polyphony, see Nettl 2010, 124.

16. For India, see Wade 2001, 24. North Indian classical music will often combine the single line of a sitar (for instance) with the percussion of the tabla, but both will share a single rhythmic pattern (*tala*). Mok 1966 notes that, though ancient Chinese ritual music was fundamentally homophonic (emphasizing octave relations), Chinese traditional music was heterophonic from very early days; see also Penyeh 1998, 2:3,11.

17. In 1920, Hugo Riemann held that "according to a majority of historians who have disputed it, rules of [polyphonic] composition are not to be found in the works of the antique writers" (Riemann 1974, xix) but "were brought into musical practice by peoples of northern ancestry overflowing southern Europe (dating from the second century before Christ)" (xx). Regarding the Greeks, Riemann mention the contrary opinion of Gevaert 1875, 2:356–357, who agreed with Wegener 1861, 3, that "the Greeks used different intervals in simultaneous harmony."

18. The surviving works and fragments are collected with authoritative transcriptions in Pöhlmann and West 2001.

19. *Republic*, 399a–c (Plato 1997, 1036).

20. *Politics*, 1342a1–2, 11–16 (Aristotle 1984, 2:2128–2129).

21. *Poetics*, 1453b10–12 (Aristotle 1984, 2:2326).

22. *Laws*, 812d (Plato 1997, 1479).

23. West 1992, 205–207 at 207, discusses the surviving evidence of heterophony in Greek music, including its mentions in Aristoxenus's important treatise.

24. *Laws* 812d–e (Plato 1997, 1479–1480). Riemann (1974, xix) thought that this passage actually proved that such embellishment "is the exact opposite of what we understand today as genuine polyphony."

25. See Barker 1990 for further discussion, as well as Crocker 2011. Despite the nineteenth-century work cited in note 17, Gentili (1990, 25) stated that Greek music "was pure melody and excluded simultaneous combinations of sound." Georgiades (1973, 23–38) even argued that the nature of Greek rhythms made polyphony "impossible," in the later Western sense; for a critique of this account of Greek rhythm, see Pöhlmann 1995.

26. The "new music" of the later fifth century BCE seems to have involved such polyphonic developments, at least a greater use of *heterophonia*; see West 1992, 103–104, 359, and Csapo 2004, 220–221, 224–225.

27. See McNeill 1995.

28. *Republic*, 580d–583a (Plato 1997, 1188–1190).

29. For a general overview of this and other ancient Greek musical notations, see West 1992, 254–276.

30. For Guido's boast, see Strunk and Treitler 1998, 215, and Kelly 2015, 61–77.

31. West 1992, 284–285; Pöhlmann and West 2001, 13, whose transcription I give in figure 1.5.

32. Nicomachus 1994, 84; see also 173, discussed in Holbrook 1983, 88.

33. *Timaeus*, 35b–37a (Plato 1997, 1239–1240).

34. *Republic*, 617b (Plato 1997, 1220). Regarding the image in figure 1.7, see Warburg 1999 and McGrath 2009, 233–238.

35. The term *harmoniai* seems to have referred to the "modes," the various octave species; see West 1992, 177–179.

36. *Republic*, 616a (Plato 1997, 1219).

37. Ibid., 617c (Plato 1997, 1220).

38. Ibid. Nor is there any suggestion of emotion in Plato's *Timaeus* as the demiurge shapes the body and soul of the cosmos "as excellent and supreme as its nature would allow" (30b) in accord with musical intervals so as produce time as "a moving image of eternity" (37d).

39. Cicero 1977, 2:42.

40. Cited and translated in the excellent article by Sullivan (1997, 36), whose translation I have followed here, except for the final sentence, taken from the version in Martianus 1977, 2:9–10. For further helpful glosses, see also Shanzer 1986, 90–95, 205–206.

41. Martianus 1977, 2:350–351.

42. Ibid., 2:352–353; the text uses the Greek term Monad (*monos*) to describe the One.

43. See, for instance, Plotinus 1992, 427–429 (*Enneads* V.1.6).

44. Martianus 1977, 2:356–357.

2   Dispassion and Deification

1. Strunk and Treitler 1998, 131.

2. McKinnon 1987, 85.

3. Ibid., 58–59; McKinnon notes that, though this treatise *De oratione* is found among the writings of Nilus of Ancyra, it is probably the work of Evagrius, who was originally from Constantinople, retired to the Egyptian desert in 382, and was "the first monastic figure to write extensively" whose writings "played an important part in the history of monastic mysticism and spirituality." It is not clear whether Evagrius's term *eurhythmōs* means "well-rhythmed" in the sense of evenness of declamation (which later became the prevalent sense of this word) or some other mode of performing the rhythms of the text.

4. Evagrius Ponticus 2003, 193.

5. Ibid.; for Evagrius's analysis of the vices addressed to beginners in the spiritual life, see, for instance, "On the Eight Thoughts," Evagrius Ponticus 2003, 66–90.

6. Ibid., 198; see also the readings in Evagrius Ponticus 1970, 63. Note that the Stoics also used the term *apatheia* to denote the state of passionless calm achieved through philosophic reflection.

7. One of them takes a melody close to the Seikilos epitaph (figure 1.4) and uses this easygoing drinking song to sing hosanna to the Messiah. A hymn to the sun becomes an antiphon to Simeon; a hymn to Nemesis takes on the words "Lord, have mercy" (*Kyrie eleison*). See Werner 1970, 338–339, 354–355.

8. Apel 1958, 301–304, at 304.

9. Hucbald, d'Arezzo, and Afflighemensis 1978, 137–138. Regarding the exact identity of this John, see 82–93.

10. "There has never been a time in the recorded history of European music—or of any music, it seems—when polyphony was unknown" (Taruskin 2005, 1:147).

11. Weiss and Taruskin 2008, 60–61; Gerald de Barri's Latinized name is Giraldus Cambriensis. For a wider discussion of popular polyphony, see Bukofzer 1940.

12. For a magisterial survey of the monophonic tradition of chant until the ninth century, see Page 2010.

13. See, for example, Holbrook 1983, 207–214, and especially the essays in Huglo 1993.

14. I thank Michel Huglo for emphasizing this to me.

15. Brown 2000, 120, situates this work as the "last contribution of a future bishop to the intellectual life of Milan," part of Augustine's project to write "textbooks of the sciences."

16. Augustine 1997, 15:347.

17. Augustine 1990, 155–157, at 155 (4.4–5). For a general discussion of Augustine's work on music, see Schueller 1988, 239–256.

18. By comparison, the decidedly anti-Christian Neoplatonist Plotinus treated the One as emanating the Dyad without mentioning any ratio between them. See Plotinus 1992, 460–463 (*Enneads* V:4).

19. Augustine 2007, 127 (*Confessions* VII:9), noted that, though in the Neoplatonists he found "that the Word, God, *was born not of flesh nor of blood, nor of the will of man or the will of the flesh*; but I did not find that *the Word became flesh*."

20. Augustine 1990, 155–157, at 155 (4.4–5). Synan 1964 suggests that a passage in Augustine's *Contra academicos* describing Plato and Aristotle "singing in unison [*sibi concinere*]" alludes to polyphony, though Holbrook 1983, 168, considers it "more likely intended to evoke the image of their ideas' fitting together into a single well-tempered mixture than of the two philosophers' singing contrapuntally."

21. From the "Commentary on Psalm 150" in Augustine 2004, 514. Cf. the possible source discussed in Synan 1964.

22. A similar argument comes forward in Augustine's short dialogue *De magistro* (*On the Teacher*) (Augustine 1995).

23. Proclus 1963, 51, proposition 52; Augustine refers to this seminal text on many occasions in *De Trinitate*.

24. Cohen (1993) admirably discusses these ancient terms in the context of his treatment of the relation between consonance and dissonance.

25. The Greeks "make a distinction that is rather obscure to me between *ousia* and *hypostasis*, so that most of our people who treat of these matters in Greek are accustomed to say *mia ousia, treis hypostaseis*, which in Latin is literally one being, three substances. But because we have grown accustomed in our usage to meaning the same thing by 'being' as 'substance,' we do not dare say one being, three substances"; Augustine 1990, 196 (5.10), emending their reference to the language of the original text from "English" to the more consistent "Latin."

26. He thinks that the Greeks, "if they like, could also say three persons, *tria prosopa*, just as they say three substances, *treis hypostaseis*. But they prefer this latter expression, because I imagine it fits the usage of their language better"; ibid., 228 (7.11). Tertullian addresses the concept of person in *Against Praxeas* (written not earlier than 208; see Roberts et al. 1885, 3:597–627, especially ch. 7, 601–602), and may have been the earliest Latin Christian author Augustine had in mind.

27. Augustine 1990, 226–227 (7.8); the scriptural citations are from 2 Cor. 2:10 and Deut. 6:4.

28. For instance, Taylor (1989, 131), argues that Augustine "introduced the inwardness of radical reflexivity and bequeathed it to the Western tradition of thought."

29. Cf. Henry 1960.

30. In Greek, *prosopon* literally means "that which is before/across (*pros*) from the eyes (*ops*)," first of all meaning the face or countenance. Homer uses this word in a plural form (*prosopa*) even when referring to a single person, implicitly indicating that even a single solitary "face" is implicitly regarded by another face, part of a larger world of faces. The word also means "one's look, countenance," which also implies some onlooker, someone "before one's eyes." *Prosopon* also came to mean "character" in the sense of a dramatic part and hence also was used for the masks in the theater that expressed the character of each personage in the drama embedded in the whole drama, not standing alone or apart from the other characters. For a reading of Plato in terms of the concept of person, see Gerson 2003. I am indebted to the insights of Venable 1993; see also Peacocke and Gillett 1987. For a general view, see Trendelenburg 1910.

31. There, the paterfamilias could give them personhood through the process of manumission, literally "releasing by hand." Because so many Roman citizens did not speak Latin, this ceremony sometimes took the form of a wordless charade: the paterfamilias would face the court, his son (or whoever was receiving manumission) facing him, hence with his back to the court. The father would then strike his son a ceremonial slap in the face that would

then turn him around to face the court, after which the son would then be a person in his own right, able to address the court. See Gibbon 1994, 2:789–790, 806–811.

32. For artificial persons in Roman law, see Borkowski 1997, 84–87; see also Duff 1971 and Watson 1967.

33. Boethius 1936, 83–85. Note that here Boethius bypasses the Roman legal tradition that allowed artificial personhood for statues and other entities.

34. See Tomlinson 1999 for a discussion of the metaphysics of voice.

35. Erickson and Palisca 1995, 13, which contains a full translation and excellent commentary on these works. For Guido's somewhat different treatment of "diaphony" or organum, see Hucbald, d'Arezzo, and Afflighemensis 1978, 77–82. For the Syrian practice of organum, see Husmann 1966; for a survey of the earliest sources, see van der Werf 1993. For general surveys of this repertoire, see Hoppin 1978, 187–200; Taruskin 2005, 1:147–156, who emphasizes the theoretical quality of perfectly parallel organum at the fifth or fourth, which would constantly run into tritones (such as B-natural versus F) and thus require introducing new notes (such as E-flat) not included in the gamut of Gregorian music (which only had the possibility of B-flat or B-natural).

36. Taruskin 2005, 1:147–148, argues this point cogently.

37. For context, see R. E. Atkinson 2010.

38. Pirrotta 1968, 284n11.

39. Erickson and Palisca 1995, 26, citing Boethius 1989, 169 (5.9). Nicomachus 1994, 173, one of Boethius's most important Greek sources, also notes that "systems [i.e., combinations of two or more intervals] are consonant when the notes comprising them, though different in compass, are, when struck simultaneously or sounded in some way, intermingled with one another in such a manner that there is produced from them a single sound like one voice."

40. Erickson and Palisca 1995, 30, translating *misceri* as "mix," for consistency of terminology (as discussed below).

41. Aristotle, *Sense and Sensibilia* (*De sensu*) 447b1–2 (Aristotle 1984, 1:709–710).

42. Erickson and Palisca 1995, 30.

43. Cohen 1993 helpfully analyzes these terms in connection with the ontology of consonance.

44. Erickson and Palisca 1995, 30.

45. See Rand 1940 and C. M. Atkinson 1999 concerning these works. In his commentary on this passage in Martianus Capella, Regino of Prüm (d. 915) explained that "*concentus* is the uniting of tones that are alike. *Succentus* occurs when different tones sound together, as, for example, in *organum*," here quoting Remigius of Auxerre (d. by 908), who adds explicitly that what Martianus described in Apollo's grove was "*duplis succentibus*, that is, in *duple organum* [*duplicis organis*]—namely, an octave and a perfect consonance." I am here indebted to Sullivan 1997, 39; for the text in question, see Remigius 1962, 87.12–13.

46. Eriugena's commentary is his *Annotationes in Marcianum*, in which he uses the verbal form of *succentus*; see Haar 1960 and Sullivan 1997. Cf. the use in the *Enchiriadis* treatises of *concentus* rather than *succentus* (see n. 42, this chap.).

47. Concerning Eriugena's musical views, see MacInnis 2014. Because Eriugena used the phrase *organicum melos*, which *Musica enchiriadis* had used in reference to polyphony, some nineteenth-century musicologists (including Hugo Riemann) took this as indicating that Eriugena was referring to contemporary organum or alternatively that the treatise drew its Neoplatonic elements from Eriugena. Yet subsequent investigations showed that the phrase *organicum melos* could also refer to nonpolyphonic practice, conceivably referring to the fixed mathematical ratios used to build instruments, organa. See Waeltner 1977. Also, the Neoplatonic references in the *Enchiriadis* treatises all could have been derived from writings of Boethius or Augustine, whom these treatises explicitly mention, whereas they never mention Eriugena. See Erickson and Palisca 1995, xliv–xlvi, and also Erickson 1992, which gives a detailed account of the controversy; some eminent musicologists, such as Michel Huglo (1988), still remain persuaded of the Eriugena connection. See Phillips 1984 for an extensive discussion and also Bower 1971, 2002.

48. For Eriugena's gloss, see Eriugena 1970 18, and his longer gloss on celestial harmony in the Oxford manuscript of his annotations, Jeauneau 1978, 123–132; for Eriugena's use of the term *coaptantur*, see his *De divisione naturæ* (*Periphyseon*), book III, chapter 6; see also Sullivan 1997, 37–39. For another interesting comparative series of glosses by Eriugena and Regino on Martianus, see Atkinson 1999.

49. These arguments are confirmed by the gloss (cited in note 42 above) from Remigius, who was a disciple of Eriugena's student Heiric of Auxerre (d. 876). Sullivan (1997, 39), considers the connection between Remigius and Eriugena to be "certain."

50. As argued in Duchez 1980.

51. *Periphyseon* (Eriugena 1987), 678c. Note that Ignatius of Antioch shortly after 1000 wrote the Church in Ephesus that "Jesus Christ is sung in your harmony and symphonic love. And each of you should join the chorus, that by being symphonic in your harmony, taking up the divine inflection together, you may sing in one voice through Jesus Christ to the Father"; this language of polyphony, though, is followed by an injunction "to be in flawless unison that you may partake of God at all times as well" (Page 2010, 41).

52. Eriugena 1987, 15 (quoting O'Meara's introduction); see also Carabine 2000, 95–102.

53. As in his "Homily on the Prologue to John's Gospel," Eriugena 1990, 55.

54. Carabine 2000, 93, observing also that "some human beings become more 'angelized' than others" (96–97); see Eriugena 1987, 179 (II, 575A), 477b (822A–B), regarding paradise as human nature.

55. Eriugena 1987, 550 (V, 883C–D), which is preceded by a set of visual exempla.

56. Ibid., 19–20.

57. Specifically, Gilbert of Poitiers, Almaric of Bena, and David of Dinant, against all of whom Thomas Aquinas later argued.

58. Eriugena 1987, 21.

59. Plotinus 1992, 709 (VI.9.10–11); I have slightly altered the translation.

60. Bett 1964, 174.

3   The Music of the Blessed

1. Dante, *Paradiso*, 1:13, showing the canto and line number; all translations from the *Divine Comedy* come from Dante 1975, giving the volume and page number in parentheses (here 3:3).

2. For an authoritative account, see Fuller 1990. The term "St. Martial organum" was a label first applied by nineteenth-century scholars.

3. Their names were recorded by an anonymous Englishman a century later who had studied in Paris and whose text became known as "Anonymous IV"; for a survey of this history, see Hoppin 1978, 200–241; Knapp 1990; Taruskin 2005, 1:169–205. Interestingly, the first extant record attesting to the use of polyphony in Notre Dame was for the Feast of the Circumcision (January 1), though Wright (1989, 338) considers that "undoubtedly" was not the first use. That date was one of the traditional occasions for the Feast of Fools, which Harris (2014) argues was not as wild or disruptive as had long been thought.

4. Note that at the word "*omnes*" the two voices begin a note-by-note texture that became known (confusingly) as "discant," as opposed to "organum" up to that point.

5. See McKinnon 1987, 153–167. For instance, Leonin's *Viderunt omnes* relies on trochaic and iambic patterns, indicated by special uses of the Gregorian neumatic notation; see Waite 1973, whose transcription I have followed in figure 3.3 and whose controversial interpretation is discussed in Page 1990a, 86.

6. For early descriptions of rests, see the anonymous *Discantus positio vulgaris* (about 1230) and especially Franco of Cologne's *Ars cantus mensurabilis* (about 1280) (Strunk and Treitler 1998, 222, 236–238), discussed in Kelly 2015, 112–141. In his *Confessions,* Augustine notices Ambrose reading in silence (6.3); Augustine's own moment of conversion happened as he "read in silence" (8.12); his final conversation with his mother considered the implications "if the soul herself were to be silent and, by not thinking of self, were to transcend self" (9.10). See Mazzeo 1962.

7. For influential interpretations of Notre Dame music, see Hilliard Ensemble 2000; Early Music Consort of London 2002; Tonus Peregrinus 2005. For the comparison of music with architecture, see Haar 1961, 401–408; for architecture as guiding metaphor, see Panofsky 1957; Binski 2010. For a critical rejoinder, see Page 1993, 1–42.

8. See Plato, *Phaedrus*, 276c, *Seventh Letter*, 344c; Aristotle, *Metaphysics*, 987b, 988a; and Szlezak 1999.

9. See Berger 2005 and Huglo 1993.

10. Wright 1989, 325–329, at 327.

11. Carruthers 2008, 182, 180, a treasury of materials bearing on the nature and implications of medieval conceptions of memory.

12. For textual polyphony in these motets, see Huot 1997, 1–55.

13. For the relation between Paris and polyphony, see Page 1990b, 137–154. For the relation to contemporary instrumental music, see Page 1986, 57–76. For the study of music in the University of Paris, see Carpenter 1972, 46–69.

14. Cited in Waite 1973, 1–2. See also Fassler 1987.

15. Torrell and Royal 2005 gives a particularly clear account of the career of Thomas in light of recent research; for his Parisian studies (1245–1248) under Albertus Magnus, see 18–35. Thomas served two periods as a master in Paris, from 1252 to 1259 (36–95) and then again from 1268 to 1272 (179–223). For the question of the audience of sacred polyphony (through ca. 1500), see Dean 1997.

16. For further consideration of the quadrivium and the relation of music to her sister arts, see Pesic 2014b, 9–20.

17. For Thomas's relation to Boethius, see Pesic 2014b, 68, 226, and McInerny 1990, 101–103. Thomas's commentary on Boethius's *On the Trinity* is one of his earliest writings (ca. 1257–1259) and shows his close study of Boethius's theological work; see Aquinas 1946.

18. *Summa Theologiae* I, Q. 29, art. 4 (hereafter *ST*; numbers in parentheses cite the translation in Aquinas 1952, here 1:166). Though he did not know Greek, Thomas considers the Greek word *prosōpon* important enough to quote verbatim, noting that it refers to masks that were "placed on the face and covered the features before the eyes" (*ST* I, Q. 29, art. 3; 1:164).

19. For a thorough study of Thomas's references to music, see Burbach 1966.

20. Wegman 2005, 19.

21. Burbach 1966, 82. The passage cited comes from Thomas's *Commentary to Psalm 32*:3; the main context seems to be the avoidance of instruments in divine service, but Thomas's reference to Aristotle's strictures against flute-playing in *Politics* 8:6 indicates a broader critique of elaborate virtuosity. For the musical views of his teacher, Albertus Magnus, see Hüschen 1970; Albertus had written a *Summa de scientia musicali* and a *Commentati musicae Boethii*, both of which are lost, as well as a commentary on Pseudo-Dionysius.

22. Aristotle, *Topics* II.10 (114b34), which Thomas cites in *ST* I Q. 12, art. 10, obj. 1; Q. 58, art. 2, obj. 1, and Q. 85, art. 5, which gives his sustained discussion of "whether we can understand many things at the same time?" (1:457–458).

23. *ST* I Q. 12, art. 4, repl. obj. 1 (1:54).

24. *ST* I, Q. 56, art. 3, 1 (1:294).

25. *ST* I Q. 57, art. 3, 4 (1:297–299).

26. *ST* I Q. 52, art. 1–3; Q. 53, art. 1–3 (1:278–284).

27. *ST* I Q. 14, art. 7 (1:81–82).

28. Weiss and Taruskin 2008, 62.

29. See Leach 2009. For the related controversy about minstrels, see Page 1990b, 8–41.

30. Aelred of Rievaulx 1990, 209–210.

31. Ibid., 210.

32. Ibid., 211.

33. Page 1990b, 3–6.

34. Wright 1989, 273–274. Michel Huglo pointed out to me that the names of the clerics in the Codex are poorly written by a different scribe in red ink on the musical staff, casting some doubt on their identities.

35. Wright 1989, 246.

36. For instance, Van Deuseun (1995, 161–176) argues that Robert Grosseteste's concept of an ordered firmament (*firmamentum*), with its ordered and overlapping strata, may have provided a theological basis for the medieval motet.

37. For a comparison between Eriugena and Thomas, see Bett 1964, 184–186.

38. Torrell and Royal 2005, 239–244, reviews all these stories, including the two cited here (241–242).

39. *Paradiso*, 10:64–138 (3:111–113).

40. Ciabattoni 2010.

41. *Inferno*, 21:139 (1:221).

42. *Purgatorio*, 2:46–8 (2:15); see also Ciabattoni 2010, 113–114.

43. *Purgatorio*, 28:59–60 (2:307); emphasis added.

44. *Paradiso*, 14:118–123 (3:161).

45. *Paradiso*, 6:121–126 (3:67).

46. See above, chapter 1.

47. This is emphasized in Ciabattoni 2010, 161–162.

48. *Paradiso*, 8:16–19 (3:83–85).

49. Regarding dances in *Paradiso*, see Ciabattoni 2010, 162–163, 187, 209, 213–214, and the discussion of liturgical dance on 151–152; see also Rokseth 1947.

50. *Paradiso*, 13:25–27 (3:143): "Lì si canto non Bacco, non Peana, / ma tre persone in divina natura, / e in una persona essa e l'umana."

51. See Cornish 2000.

52. *Paradiso*, 10:139–148 (3:117).

53. Thorndike 1941.

4   Oresme and the "New Song"

1. See Hayburn 1979, 17–23, at 20–21; for an overview of papal decrees regarding polyphony, see 392–394.

2. This is the judgment given in Hayburn 1979, 21–22.

3. Ibid., 20–21.

4. Ibid., 21.

5. Ibid., 22.

6. For a useful overview of the history of this and related terms, see Fuller 2002; Wegman 2014.

7. For the larger context, see Pesic 2014b, chap. 2.

8. See Pesic 2014a.

9. Oresme 1971, pt. 3.

10. See Pesic 2003, 5–21.

11. The fundamental reasoning is simple: if two planets' speeds are in the ratio of integers m:n, then after $m \times n$ revolutions they will have "lapped" each other a sufficient number of times so as to have regained their initial positions. See Pesic 2014b, chap. 2.

12. Oresme 1971, 316–317.

13. Anne Sexton noted her "secret instructions to herself as a poet" to include "whatever you do, don't be boring" (Middlebrook 1992, 158).

14. Oresme 1968a, 480–481.

15. For the "new song," see Rev. 5:9, 14:3 and Ps. 39:4, 143:9, 149:1.

16. Oresme 1968a, 480–483.

17. For a discussion of Oresme's consideration of whether geocentric astronomy is correct, see Pesic 2014b, chap. 2.

18. See Page 1993, 30–42.

19. Rev. 5:9–10 (Douai-Rheims 1899 translation used throughout).

20. For a medieval fiddler preparing to play a "new song," see Page 1990b, 1.

21. Rev. 14:1–5.

22. Rev. 21:5; 2 Cor. 5:17.

23. Oresme 1968a, 482–483.

24. For the controversy, see Page 1986, 77–84; Page 1990b, 110–133. Regarding the *carole,* see Mullally 2011.

25. Oresme 1968a, 482–483.

26. Ibid., 484–485, 482–483, 484–485.

27. Oresme 1968b, 164–167. Oresme's work on configurations here seems to develop from an earlier tradition (see the anonymous *Tractatus bonus de uniformi et difformi*, cited here, 576–621) and from Oresme's own commentary on Euclid's *Elements*, cited here 526–575 and given as a whole in Oresme 1961.

28. Oresme 1968b, 168–169.

29. Ibid., 174–175.

30. Ibid., 176–177.

31. Ibid., 304–305. For Aristotle's discussion of this problem of intermittency, see *Sense and Sensibilia* (*De sensu*) 448a19–b12 (Aristotle 1984, 1:711).

32. Oresme 1968b, 304–305.

33. Ibid., 306–307.

34. Babbitt 1985.

35. Oresme 1968b, 310–311.

36. Ibid., 310–313.

37. Ibid.

38. This confirms the connection between irrationality and dissonance in Pesic 2014b, 17.

39. Oresme 1968b, 314–317. Boethius had made a similar comparison, as cited in on 307n4 from his *Institutes*, IV.5. Curiously, his example of a color top was precisely the same experiment that the young James Clerk Maxwell used as he began his work on color theory.

40. See Pesic 2014b, chap. 2.

41. See Hoppin 1978, 353–357; Taruskin 2005, 1:247–255; for de Vitry's innovations in notation, see Kelly 2015, 142–175. For the connection between music and mathematics, see Page 1993, 112–139.

42. Pesic 2014b, chap. 2.

43. Oresme 1968b, 314–317.

44. Ibid., 314–317.

45. Ibid., 322–323.

46. Ibid., 324–325.

47. Ibid., 326–327.

48. Ibid.

49. Ibid., 334–335.

50. Ibid., 334–337, 338–339; concerning his critique of miracles and divination, see Oresme 1985, 17–25, 50–73.

51. Oresme 1968b, 338–339, 342–345.

52. Ibid., 316–317. Cf. Descartes's telling of this same story; see Clark and Rehding 2001, 1; Pesic 2014b, chap. 6.

53. Oresme 1968b, 334–335.

5 Polyphonic Controversies

1. Wegman 2005. See also the controversy about listening to polyphony addressed by Dean 1997; Meconi 1998.

2. For the case of the Cistercians, see Page 1990b, 155–170; Leitmeir 2007.

3. Wegman 2005, 19–20.

4. Ibid., 18.

5. Wright 1989, 347–348, 371.

6. Wegman 2005, 19.

7. Ibid., 20.

8. Ibid., 21–22.

9. For a helpful overview, see Macey 1998, 11–31, which discusses the expulsion of the Jews and the laws against sodomy on 16–17.

10. Wegman 2005, 37, 201n51.

11. Ibid., 57.

12. Macey 1998, 17–19.

13. Wegman 2005, 43–44.

14. Macey 1998, 96.

15. Macey 1998, 74–75.

16. For an extensive discussion and recorded examples, see Macey 1998, 98–117; see also Macey 1999.

17. Macey 1998, 92.

18. Wegman 2005, 1–9.

19. Ibid., 9–15.

20. Ibid., 109.

21. Ibid., 163–165.

22. Ibid., 133–147.

23. Ibid., 183.

24. Ibid., 43.

25. Paul IV's statement was in 1557; see Macey 1998, 229.

26. Wegman 2005, 80–81.

27. Ibid., 82.

28. Castiglione 2002, 55.

29. Ibid., 56–57.

30. The incident is recounted and the extant account translated in Wegman 2005, 92–95.

31. Ibid., 93–94.

32. Ibid., 53.

33. Ibid., 63–64.

34. For the quadrivium, see Pesic 2014b, 19–22.

35. Vasari's description of Verrocchio's gifts is from Winternitz 1982, 3.

36. Winternitz 1982, xxii–xxiii. For the conjecture that Leonardo appears as Orpheus in figure 5.2, see Duffin 2015.

37. See Winternitz 1982, 5–16.

38. Gaffurius 1969, 123.

39. For Leonardo's participation in *feste* during his Milan years, see Winternitz 1982, 73–93, cited on 76.

40. For Leonardo's work on instrument design, see Winternitz 1982, 137–167, cited on 138.

41. Ibid., 168–186, esp. 180–181.

42. The work in this form was only published in 1651. For the sections regarding music, see Winternitz 1982, 204–223.

43. Ibid., 215. I have modified Winternitz's translations so that they are closer to the original language.

44. Ibid., 219. For Alberti in context, see Grafton 2000, 115–149.

45. Winternitz 1982, 216–217.

46. In contrast, Winternitz reads Leonardo's description as an "absurd picture" (ibid., 218), evidently not thinking of the realities of part-books.

47. Ibid., 205.

48. Ibid., 206n4.

49. Ibid., 212.

50. Here he may have been influenced by Gaffurius (1969, 69), who notes that "rhythm, in the opinion of Aristides, consists of units of time in space."

51. Winternitz 1982, 216–218.

52. Ibid., 215–216.

53. The myth has been traced back to 1607; for relevant documents, see Palestrina 1975. For an authoritative discussion of what really happened at the Council of Trent, see Monson 2002; an older, more incomplete account is Hayburn 1979, 25–31.

54. Monson 2002, 16.

55. Loyola 1992, 38.

6 *E pluribus unum*

1. Wegman 2005, 65.

2. For a thoughtful consideration of Tinctoris's list, see Wegman 1995.

3. For an excellent, detailed, and accessible treatment of Machaut's *Messe de Nostre Dame*, see Leech-Wilkinson 1990.

4. For an account of this modern revival, see Leech-Wilkinson 2002. For the detailed context in the traditions of Reims, see Robertson 2002, 257–275.

5. The quote is from Franco of Cologne; Strunk and Treitler 1998, 229.

6. Figure 6.3 comes from Bede's treatises on time and cosmology (about 703 CE); the passages about the planets are translated in Bede 2010, 80–84.

7. The Kyrie III has a tenor isorhythm alternating ten with seven notes; Leech-Wilkinson 1990, 27, suggests that "all the Kyrie *taleae* are deliberately related."

8. He does use isorhythm, though, for the concluding "Amen" at the end of the Credo; ibid., 43–45.

9. Ibid., 29–45.

10. See Reese 1959, 238, and Blackburn 2000, 76–82; Haar 1976 gives a careful account of the possibilities and of the use of this subject in Josquin's mass.

7 Polyphony and Power

1. This and the following paragraph draw from Moroney 2007, which also tells how these works were rediscovered after long having been considered as lost.

2. At St. Mark's in Venice, Adrian Willaert had written such polychoral works already in the 1540s, which Giovanni Gabrielli and others continued in the 1580s.

3. See Moroney 2007, 28–33, at 30.

4. Quoted from the artist's statements in *Janet Cardiff's 40 Part Motet* 2015, 3′22″–3′45″.

5. Cardiff and Miller 2015.

6. Janet Cardiff: Forty-Part Motet 2015.

7. *Janet Cardiff's 40 Part Motet* 2015.

8. Dwyer 2013.

9. Horn 2013.

8   Controlling Dissonance

1. G. Galilei 2000, 107.

2. Mersenne 1963, 3:211. See also Kursell 2013, 196–197.

3. Pesic 2014b, 89–102.

4. Descartes 1996, 1:21.

5. Descartes 1970, 1:340.

6. Regarding the *sensorium commune*, see the interesting discussion in Heller-Roazen 2009, 31–42.

7. Harrington 1987, 6–7; Finger 2000, 69–83.

8. Weiss and Taruskin 2008, 2:142.

9. For a classic discussion of Palestrina's contrapuntal technique and control of dissonance, see Jeppesen 2012.

10. I remember Philip LeCuyer pointing this out in the early 1980s during music tutorial meetings at St. John's College in Santa Fe.

11. Consider, for example, the 1585 presentation of a version of Sophocles's *Edipo Tiranno* (*Oedipus the King*) at the Teatro Olimpico in Vicenza with choruses by Andrea Gabrieli; see Schrade 1960. The term "opera," however, was not consistently used until the nineteenth century; Abbate and Parker 2012, 37–44, gives a helpful corrective to older, simpler narratives of the "birth of opera."

12. For other aspects of Vincenzo Galilei's projects, see Pesic 2014b, 48–54.

13. According to Claude Palisca's preface to V. Galilei 2003, xvii.

14. V. Galilei 2003, 222.

15. See Zarlino 1968, 2011; Vincenzo and Zarlino engaged in a protracted polemic.

16. V. Galilei 2003, 212.

17. Ibid., 208.

18. Ibid. 224.

19. Ibid., 362–363.

20. Ibid., 225.

21. Ibid., 224–225.

22. See Warburg 1999; for the music, see Walker 1963, described on xi–xxxi.

23. As noted by Walker 1963, xvii: "The multiplicity of parts end by canceling themselves musically. In fact, the ear finally loses the contour and details of such a numerous mass of sounds."

24. Strunk and Treitler 1998, 536–544, on 539–540.

25. Ibid., 540.

26. Caspar 1993, 248; see also Gingerich 1993, 396–398.

27. Pesic 2014b, 73–88.

28. Kepler 1997, 243.

29. Ibid., 440.

30. See Pesic 2014b, 82–85.

31. Kepler 1997, 411.

32. Westfall 1980, 510n136, quoting Newton's niece Catherine Barton, who told this to her husband John Conduitt.

9   Contrapuntal Science and Art

1. Gaffurius 1993, 39.

2. For Guido's "principles of organum," see Hucbald, d'Arezzo, and Afflighemensis 1978, 77–82. For a helpful survey of medieval treatments, see Fuller 2002.

3. Christensen 2002, 2–8, at 5n12; this whole passage gives a very helpful overview of the use of *theoria* in this context.

4. Morley 1973, v.

5. For the study of counterpoint in the universities, see Carpenter 1972, 59, 139, 288.

6. Tinctoris 1961, 140.

7. In his 1473–74 *Proportionale musices* (see Strunk and Treitler 1998, 292–293).

8. Knud Jeppesen, quoted in Palisca 1985, 8.

9. For *theorica* in the astronomical context, see Pederson 1981.

10. Gaffurius 1993, 5.

11. Gaffurius 1969, xv.

12. He devotes the whole fourth book to these matters: ibid., 165–267, commenting on rational and irrational proportions on 165–167.

13. Ibid., 131–137.

14. Ibid., 160–161.

15. Ibid., 161.

16. Ibid., 266.

17. Zarlino 1968, 264.

18. Corwin 2008, 110, the only currently available translation of Part I of Zarlino's *Le Istitutioni harmonice*.

19. Ibid., 109.

20. Ibid. 109, 113.

21. Ibid., 121.

22. Ibid., 123.

23. Ibid., 130–131.

24. Ibid., 135.

25. Ibid., 175.

26. Ibid., 254.

27. Zarlino 1968, 60–61.

28. Zarlino 1968, 52.

29. Ironically, Willaert himself had composed some highly chromatic works of which Zarlino seemed unaware; see McKinney 2016, 5–11. Regarding Vicentino's innovations, see Pesic 2014b, 60–72.

30. Zarlino 1968, 288.

31. Ibid., 290.

32. Morley 1973, 9, which modernizes the spelling.

33. Ibid., 10.

34. Ibid., 140.

35. Ibid., 147.

36. Bent 2002, 595.

37. Fux 1965, 17.

38. Fux 1725, 1.

39. See Fux 1965, 22n3.

40. Ibid., 19.

41. For the context of Fux's work, see Edler and Riedel 1996.

42. As noted by Bent 2002, 566–568. In *Il Transilvano* (1610), Girolamo Diruta presented polyphonic composition as a series of types beginning with note against note; there is no evidence Fux was aware of this book. Subsequent generations learned species counterpoint from Fux.

43. Bent 2002, 565–568.

44. Fux 1965, 49.

45. Ibid., 55–56.

46. Ibid., 56.

47. Ibid., 60n2. See Roth 1926, 89; Jeppesen 2013, 138.

48. Fux 1965, 64.

49. Ibid., 71.

50. Ibid., 112.

10   In Bach's Hands

1. Webern 1963, 34.

2. Wolff 2000, 1–11.

3. Ibid., 9.

4. Ibid., 335–339, argues similarly that the simpler canon BWV 1072, which Bach entitled *Trias harmonica* (the harmonic triad), contrapuntally creates that triad.

5. Spitta 1979, 125, as per the translation in Fux 1965, xv.

6. Fux 1742, 1–2n1. Regarding Mizler's intellectual and philosophical activities, see Birke 1966, 67–82; Felbick 2012. For his commentary on Fux, see Federhofer 1996.

7. Wolff, Mendel, and David 1999, 399.

8. Ibid., 323.

9. These notes are helpfully collected in J. S. Bach 1995, which in turn was based on Niedt 1989.

10. Bach's signature B–A–C–H is also found in the tenor voice of measure 49, Variation IV. For his enormous hand span, see Wolff, Mendel, and David 1999, 369.

11. Ibid., 361.

12. C. P. E. Bach 1949, 42. See also Troeger 2003, 207–217.

13. Lindley 1993, 12, 26 (a superb resource of examples and images, along with practical suggestions about playing with the original fingerings). For more general surveys (including variant traditions between various nations), see Ferguson 1975, 67–84; Lindley 1989, 1990, 1992; Johnson 1994, 26–31.

14. Lindley 1993, 16, for Nikolaus Ammerbach's rules and exercise of 1571; Erbach used the numbering 1 2 3 4 5 6 7 8 9 10 for the modern 5 4 3 2 1 1 2 3 4 5 (Lindley 1993, 26). For the fingerings of Bach's predecessor Jan Sweelinck and Heinrich Scheidemann, see Dirksen 2007, 155–167; van Dijk 2002.

15. Godman 1957.

16. C. P. E. Bach 1949, 43.

17. See, for instance, the discussion in Kosovske 2011, 111–135.

18. Couperin 1717, 29. Regarding the connection with Bach, see Moroney 2003.

19. Wolff, Mendel, and David 1999, 434; Forkel's biography was first published in 1802.

20. Ibid.

21. Wolff 2000, 138, though "without drawing a parallel" between these two cases. For an account of fingering by 1789, see Türk 1982, 127–189.

22. Wilhelm Friedemann was said to be able to play right-hand scales very rapidly using the 3 4 3 4 fingering, which Mattheson still recommended in 1735; see C. P. E. Bach 1949, 69n20, and Türk 1982, 146.

23. Ibid., 69.

24. Galilei 2003, 204, 218.

25. Wolff, Mendel, and David 1999, 320.

26. Ibid., 397.

27. "Ricercar" is an elaborate fugal form Vincenzo Galilei called "particularly mischievous" (Galilei 2003, 218). For the theories of fugue before Bach, see Walker 2004; for the implications of the term "ricercar" in Bach's time, see Wolff 1991, 328–330.

28. As stated by Friedrich Wilhelm Marpurg in 1760; Wolff, Mendel, and David 1999, 363.

29. For the pervasive importance of dance in Bach's music, see Little and Jenne 2001.

30. Spinoza 2000, 125, scholium to Proposition 13, Part Two. He also argues that "an emotion which is related to several different causes, which the mind contemplates simultaneously with the emotion itself, is less harmful and we suffer less through it and are less affected towards each cause, than another equally great emotion that is related to fewer causes, or to one cause alone," because "an emotion is bad, i.e. harmful, only in so far as the mind is hindered by it from being able to think" (295, Proposition 9, Part Five).

31. For Bach's possible relation to Spinoza, see Butt 1997.

32. Wolff, Mendel, and David 1999, 455. Cf. the account of learning to play jazz via hand positions in Sudnow 2001.

33. For a detailed discussion, see Butler 1977, 64; Bonds 1991, with critical commentary in Williams 1983; Walker 2004; Troeger 2003, 227–250, gives a brief overview.

34. Butler 1977, 72.

35. Wolff, Mendel, and David 1999, 455.

36. Quoted in Margulis 2014, 18.

37. Butler 1977, 76.

38. Schmidt calls the follower an *aeteologia* that explicates the origin of the theme (*proposition*); ibid., 68–77 at 76.

39. Ibid., 84.

40. Bacon 1985, 132.

41. Butler 1977, 97–98.

42. Ibid., 98.

11   Polyphony Extended

1. Schoenberg used this phrase in a 1930 letter about Mendelssohn, Wagner, and Brahms; see Dahlhaus 1987, 182.

2. Swafford 1997, 262.

3. By comparison, African rhythms treat 3 against 2 "as being just as normal as beating in synchrony" (Chernoff 1981, 94).

4. See Brinkmann 1999, 146–147.

5. Helmholtz 1954, 236. Regarding Helmholtz's approach to sound and instruments, see Hui 2013, 55–87.

6. For further discussion, see Pesic 2014b, 223–228, for the larger history of sirens, see Rehding 2013.

7. See Kursell 2013. Schoenberg does not refer to Helmholtz anywhere in his writings, but Dahlhaus 1987, 142–143, considers them directly linked, as does Steege 2012, 220, 235–240. Helmholtz's work was so much in the air during that period, especially among educated musicians, that it is hard to believe that his ideas were unknown to Schoenberg. For instance, Mahler (with whom Schoenberg was in close contact) was very interested in Helmholtz; see Johnson 1999, 251n32.

8. For timbre and Haydn's development of the orchestra, see Dolan 2013.

9. See Dahlhaus 1987, 141–143.

10. For a helpful commentary on Webern's orchestration, see Dahlhaus 1987, 181–191.

11. Webern 1963, 19.

12. Ibid., 18.

13. For instance, see Reti 1978.

14. Ligeti 1983, 25.

15. For a helpful discussion, see Maconie 1990, 86–93.

16. Ligeti 1983, 14–15; see also Clendinning 1989; Steinitz 2003, 91–93 (*Gruppen*), 103–111.

17. Ligeti 1983, 14. Regarding this work, see Steinitz 2003, 96–113.

18. In *Zeitmasse* for five wind instruments (1955–56), in measures 201–205, Stockhausen had individual instruments play at different tempi and complex rhythms, though performances often involve a conductor giving a larger stabilizing beat.

19. For the dating of these studies, see Gann 1995, 69; for transcendental numbers, see Pesic 2003, 62, 66, 150.

20. Gann 1995, 184–188, further analyzes this piece.

21. For a detailed discussion of the string quartet arrangement and the Arditti's rendition, see Callendar 2016.

12   Contrapuntal Radio and Polyphonic Fields

1. For Gould's radio documentaries, see Kostelanetz 1988; Canning 1989; Tulk 2000; Plowright 2008; Mauer 2010.

2. Gould 1984, 374–388, on 379; 391–394.

3. For Gould in the larger context of Canada and the idea of the North, see McNeilly 1996; Grace 2015, 12–14, 130–136; Davidson 2005, 7, 131–132; Neumann 2011.

4. For a textual version of this prologue, see Gould 1984, 389–390.

5. Ibid., 377.

6. In 1962, Gould and the actor Claude Rains recorded Richard Strauss's melodrama *Enoch Arden* (1897); see Gould 1984, 100–103.

7. For Gould's commentary on this work, see Gould 1984, 377–388, 394–395.

8. Kierkegaard 1964.

9. Gould 1984, 415–416.

10. For the biographical background of Cage's turn from music as expression to its use "to sober and quiet the mind," see Revill 1992, 84–93, on 90.

11. Kostelanetz 2003, 230.

12. Ibid.; Revill 1992, 79.

13. Ibid., 80; Cage 1969, 133.

14. The four texts are presented in Cage 2011, 194–259.

15. See Bernstein 2014 for discussion of this work.

16. Cage 2011, 194.

17. Ibid., 194–195. For Roman Jakobson's discussion of the linguistic problem of pronouncing two elements at once, see Heller-Roazen 2013, 140–141, which treats "contrapuntal recitative" on 164–165.

18. For his radio compositions, see Kostelanetz 1988; 2003, 155–167; Grassmann and Weiss 1988; Silverman 2012, 29–30, 138, 235.

19. Kostelanetz 2003, 70.

20. For an extensive discussion, see Gann 2010 and also Shultis 1995; Kahn 1997; Kostelanetz 2003, 65–67; Kania 2010.

21. Kostelanetz 2003, 245.

22. Cage 1969, 134.

23. Kant 1965, 152–155.

24. See Byrd 1993; Hunchuk and Dasilva 1989, 1990; Gresser 1998; Silverman 2012, 189.

25. Pritchett 1996, 155–156, commenting on the recorded performance at 212n23.

26. Regarding improvisation, see Kostelanetz 2003, 229–231.

27. Helmholtz 1954, 63–65.

28. Merce Cunningham: Dance Capsules 2015.

29. For Cage's relation to dance, see Kostelanetz 2003, 193–200. Cage's dedication came about because of a lecture appearance he made at Long Beach City College in 1963, which had been arranged by Fred Mauk and me, then both teenagers passionate for new music. I take my appearance as dedicatee to be yet another chance event, with the significance of all such events.

30. Ibid., 1.

31. Ibid., 191.

32. Duncan's slow dance is described in Daly 2010, 200.

13 Polyphonic Selves

1. Whitehead 1967, 82.

2. Ellenberger 1970, 145.

3. Aristotle treats the "common sense" in *Sense and Sensibilia* (*De sensu*) 449a3–34 (Aristotle 1984, 1:712–3).

4. Young 1970, 48–49; Finger 2000, 119–136; for the relation of Gall's work to his musical background, see Eling, Finger, and Whitaker 2015.

5. Harrington 1987, 8–9.

6. La Mettrie 1912, 120.

7. Harrington 1987, 18–19.

8. Ibid., 21.

9. Ibid., 206–234.

10. As suggested in 1868 by a Dr. Jensen; see Harrington 1987, 121–122.

11. Ibid., 44–45; see also Finger 2000, 137–154.

12. This "law of symmetry" was associated with François Xavier Bichat; see Harrington 1987, 15–17, 51–57.

13. Ibid., 108–109; the case was widely known in France as *la dame de Macnish*.

14. The spiritualistic background is emphasized by Brown 1983.

15. Goethe 2008, 35–36 (ll. 1112–1113); Stevenson 1991, 43.

16. Luys 1879, 551.

17. Ibid., 533.

18. Ibid., 283.

19. Wernicke 1977, 94, which also refers to piano playing on 93, 116.

20. On his own piano playing, see Wernicke 1977, 14.

21. Harrington 1987, 130–135, at 130; see also Herron 2012, 47–51, for an account of this movement.

22. Harrington 1987, 132.

23. J. Jackson 1905, 225.

24. Crighton-Browne 1907, 652; J. Jackson 1905, 190, capitalization in original.

25. J. Jackson 1905, 189.

26. Ibid., 192.

27. Ibid.

28. For an excellent account of Hughling Jackson's role in this history, see Harrington 1987, 206–234. Note that he is no relation to the John Jackson discussed above, the advocate of ambidextral training.

29. See Lorch and Greenblatt 2015; Johnson and Graziano 2015.

30. J. H. Jackson 1866, 175.

31. Glymour 1991.

32. For Gestalt psychology, see Harrington 1996, 103–139.

33. Goldstein 2005, 3–6, helpfully distinguishes the "vertical fragmentation" that Freudian psychoanalysis applied to the self from the earlier "segmentation" between different roles within the self and the "horizontal fragmentation" of the self considered as an accumulation of discrete pieces.

34. From "The Moses of Michelangelo (1914)" in Freud 1963, 80–81.

35. Lashley 1963, 155–156; Harrington 1987, 268–270, gives the background of the rise of behaviorism.

36. Harrington 1987, 270.

37. Penfield and Jasper 1954; Penfield 1975.

38. Penfield 1975, 11. The metaphor of a symphony is also used by Eagleman 2015, 42, 160.

39. Finger 2000, 281–300.

40. Harrington 1987, 22, 129, 152–154; the quote is from Camille Sabatier (129). Sacks 2008, 100, points out that the corpus callosum of musicians is enlarged, corresponding to its function of linking together the hemispheres.

41. Sperry 1968, 727.

42. See Kluft and Fine 1993; Michelson and Ray 2013.

43. Radkau 2009, 372.

44. For the relation between Helmholtz and Weber, see Radkau 2009, 370–373.

45. Weber 1958, 67–68; 1972, 51.

46. Weber 1958, 76; 1972, 58.

47. Weber 1958, 68; 1972, 51.

48. Weber 1958, 68; 1972, 51–52.

49. Weber 1958, 69; 1972, 52.

50. Angèle Christin directed my attention to the distinction in Durkheim 2014 between "mechanical solidarity" (homogeneity and imitation) and "organic" (division of labor and collaboration between different functions) forms of social order that may parallel homophony and polyphony, respectively.

51. Proust 2005a, 59.

52. Ibid., 68; the ensuing discussion of sound (68–71) is particularly important.

53. Proust 2005b, 42–43.

54. Proust 2003b, 362–363.

55. Ibid., 363.

56. See Proust 2003a, 272; 2005b, 259.

57. Proust 2003b, 5–6.

58. Ibid., 7.

59. Ibid., 391.

60. Ibid.

61. Ibid., 393.

62. Thacker 2004, 104–119, at 106. For other linguistic and poetic invocations of polyphony, see Heller-Roazen 2013, 140–141, 164–165.

63. Pound 1918, 98–99.

64. Lowell 1916, 331.

65. Bakhtin 1984, 2.

66. Ibid., 67. Hermans 1999 draws on Bakhtin to describe "a multivoiced and dialogical self" from "the polyphony of the mind."

67. Festinger 1962, 1–31.

68. For instance, see Festinger 1964, 8–32.

69. Festinger 1956.

70. Bourdieu and Abdelmalek 1964, 102.

71. Bourdieu 1986, 173–174. For Edmund Husserl's argument on the fundamental unity of subjectivity, see Husserl 1970, 1:66–67.

72. Goffman 1974, 299.

73. Lahire 2011, 31.

74. Ibid., 32.

14   Tuning the Brain

1. For a helpful summary of the elements of this consensus, see Dehaene and Changeux 2011.

2. This notion of the "Cartesian Theater" has been forcibly controverted, among others, by Dennett 1991, 101–138; for the homunculus, see 14–15, 259–262.

3. See the case history of such damage to V1 in Sacks 1995, 3–41.

4. For an overview, see Finger 2000, 197–216.

5. Figure 14.1 is taken from Rámon y Cajal 1954, 76, his last book (first published in 1935).

6. For the work of Keith Lucas and Edgar Adrian that established this, see Finger 2000, 239–258.

7. Yuste 2015, 487.

8. Ibid., 489–490; Lorente de Nó 1933, 1934, 1951; see also Larriva-Sahd 2014.

9. Buzsáki 2006, 158–159, describing Hebb 1949, which discusses Lorente de Nó on 61–66, 72–74.

10. Hebb 1949, 62; italics in the original.

11. Buzsáki 2006, 159.

12. Ibid.

13. Hebb 1949, xvii; for the early development of connectionism, see Boden 2006, 1:268–278. For Turing's relation to this work, see Teuscher 2012.

14. Minsky 1986 discusses "a scheme in which each mind is made of many smaller processes" called "agents."

15. Caton 1875.

16. See Swartz 1998. Berger 1969 collects and translates his original papers.

17. Lucier and Stone 2012, 65–78.

18. See the comprehensive text, Niedermeyer and da Silva 2005.

19. For predators and early humankind, see Mithen 2006, 132–133, 213.

20. Nietzsche 1997, 143, aphorism 250.

21. Wiener 1965, 199–202, on 199, a chapter not included in the original 1948 edition of that book.

22. As Wiener emphasized (1965, 199), this requires that there be some finite nonlinear component connecting the oscillators.

23. Gray et al. 1989, 336; Singer 2001, 2005.

24. Crick and Koch 1990, cited from Crick 1994, 245; italics in the original. See also Stryker 1989; Yamanoue 1996; Koch and Crick 2003; for a critical response, see Hardcastle 1996.

25. Crick 1994, 249.

26. Llinás 2001, 126; italics in the original.

27. Ibid., 128. For another approach to the binding problem via "the multiplicity of the brain," see Lancaster 1999, as well as many other papers in Rowan and Cooper 1999.

28. For a helpful edition of Turing's original paper, see Petzold 2008.

29. Von Neumann 2000, 51.

30. Llinás 2001, 124.

31. For comprehensive surveys, see McClelland, Rumelhart, and PDP Research Group 1986; Morris 1989; Arbib 2003. For the concept of convergence zones in neural processing, see Damasio 2000, 332–335.

32. For a detailed exposition, see Smolensky 1986.

33. For the history of connectionism, see Boden 2006, 2:883–1001.

34. Crick 1989, discussed in Boden 2006, 2:953–955.

35. See, for instance, Plomp and Levelt 1965.

36. McNeil 2007.

37. Another example of this U-shaped curve is found in Bärbel Infelder and Jean Piaget's observations of the ways children learn the balancing of two different weights at various distances on either side of a fulcrum, which McClelland 1989 modeled by an artificial neural network.

38. James 1950, 1:258.

39. Ibid., 1:259–260.

40. Ibid., 1:257–258. Cf. the assertion of Sacks 2008, 319, that "there is only a single sentence" devoted to music in James's book.

41. James 1950, 1:255.

42. Sacks 2008, 121–128, at 122.

43. See Gazzaley and Rosen 2016, 95–96.

44. Brown 2006, 31, quotes a teenager with ADHD who experienced "four different [TV] stations coming in on one channel"; the clinical study quoted is Banich et al. 2009.

45. Holton 1986, 233, from his classic essay "Metaphors in Science and Education," 229–252.

46. Ibid., 232. See also Pesic 2014b, 161–179.

47. Holton 1986, 235.

48. Ibid.

49. Von Neumann 2000, 51.

15   Music of the Hemispheres

1. Buzsáki 2006, which received glowing reviews in *Science* (Fries 2006) and *Nature* (Mehta 2007), among other scientific journals.

2. Buzsáki Lab 2016.

3. Buzsáki 2006, 136–174, reviews the neurophysiology of these oscillations and notes that, "to date, general rules are exceptional for brain oscillations. The coupling properties of the rhythmic networks should be determined experimentally in each and every cases, aided by computer modeling" (173).

4. Ibid., 23.

5. Llinás 1988.

6. Buzsáki 2006, 23–24.

7. Ibid., 146–147.

8. Ibid., 147n13.

9. Ibid., 148.

10. Ibid., 97, 97n21.

11. Ibid., 98.

12. Ibid., 153–154.

13. Ibid., 61, which surveys inhibition on 61–79.

14. Ibid. 112–113; Penttonen and Buzsáki 2003.

15. Buzsáki 2006, 115, 117.

16. Ibid., 115n9, quoting James 1950, 609.

17. Buzsáki 2006, 115–116.

18. Watson and Buzsáki 2015a, 74.

19. Ibid.

20. Lee and Wilson 2002, as summarized by Watson and Buzsáki 2015a, 74.

21. Girardeau et al. 2009. This work has been further confirmed and extended in Maingret et al. 2016.

22. Watson and Buzsáki 2015a, 75.

23. Ibid., 78.

24. Ibid.

25. Buzsáki, Logothetis, and Singer 2013.

26. Ibid., 753.

27. Ibid., 757.

28. Ibid., 757–758.

29. Ibid., 758–759.

30. Buzsáki 2006, 199.

31. Pfurtscheller and Berghold 1989, discussed by Buzsáki 2006, 199.

32. LaFleur et al. 2013; Scherer and Pfurtscheller 2013.

33. Buzsáki 2006, 229–230.

34. Ibid., 222, treating this whole process on 220–230.

35. Ibid., 221.

36. Ibid., 263–268.

37. Ibid., 244.

38. Ibid., 234–238, discusses the various views.

39. Ibid., 237.

40. Koch et al. 2016, 307, 313, 317.

41. Ibid., 317.

42. Buzsáki, private communication, 2016.

43. Buzsáki 2006, 247.

44. Buzsáki 2010; Watson and Buzsáki 2015b.

45. Buzsáki 2010, 379.

46. Changeux 2004, 87–95 (neuronal workspace), 105–110 (neuronal melody), quoted on 124, 243.

47. Cage 1969, 135, quoting the composer Christian Wolff.

48. Grinvald et al. 2003, 423.

49. Ibid., 426–427.

50. For examples of popularizing expositions, see Calvin 1990; D'Ardenne 2013.

51. Dickinson and Nagy 1983; Nagy and Dickinson 1983.

52. Frégnac et al. 2010, 172.

53. In British musical terminology, "leader" specifically means the concertmaster; I assume that Frégnac et al. (2010) mean "section leader," which would make more sense in their context.

54. There is no indication that these cells also have the orchestral section leader's job to "mark up" the section's parts with bowings.

55. For contemporary treatments of the processing of music by the brain (which I have not attempted to treat in this book), see Peretz and Zattore 2003; Levitin 2007; Koelsch 2013.

56. Janata, Tillmann, and Bharucha 2002, 121.

57. See Litweiler 1984; Jost 1994; Jenkins 2004.

58. Mazzola and Cherlin 2009.

59. Ligeti himself was deeply influenced in his own polyphonic development by Pygmy and African music; see Duchesneau and Marx 2011.

Postlude

1. Adams 2015 is the source of this and the succeeding quotes from the composer of *Sila*.

2. For a discussion of music and "embodied cognition," see Leman 2008.

3. Koenig (1991) discusses his discovery of what he called *Bewegungsfarbe*. See also Steinitz 2003, 165; Ligeti tried to reproduce this effect in his harpsichord composition *Continuum* (1968).

4. "World's Fastest Drummer: Tom Grosset Strikes Drum Over 1,200 Times in 60 Seconds" 2016.

5. Himebauch 2016.

6. Canales 2015 gives a very interesting account of the preceding twentieth-century controversy between physical and philosophical ideas of time.

7. See, for instance, Karmarkar and Buonomano 2007; Goel and Buonomano 2014.

References

Abbate, Carolyn. 2004. Music—drastic or gnostic? *Critical Inquiry* 30:505–536.

Abbate, Carolyn, and Roger Parker. 2012. *A History of Opera.* New York: W. W. Norton.

Adams, John Luther. 2015. *Sila: The Breath of the World (2014).* Fairbanks, AK: Taiga Press.

Adler, Guido. 1886. *Die Wiederholung und Nachahmung in der Mehrstimmigkeit: Studie zur Geschichte der Harmonie.* Leipzig: Breitkopf & Härtel.

Adler, Guido. 1908. Über Heterophonie. *Jahrbuch der Musikbibliothek Peters* 15:17–27.

Aelred of Rievaulx. 1990. *The Mirror of Charity.* Kalamazoo, MI: Cistercian Publications.

Agawu, V. Kofi. 1995. *African Rhythm: A Northern Ewe Perspective.* Cambridge: Cambridge University Press.

Allport, D. Alan, Barbara Antonis, and Patricia Reynolds. 1972. On the division of attention: A disproof of the single channel hypothesis. *Quarterly Journal of Experimental Psychology* 24:225–235.

Apel, Willi. 1958. *Gregorian Chant.* Bloomington: Indiana University Press.

Aquinas, Thomas. 1946. *The Trinity and The Unicity of the Intellect.* Trans. Rose Emmanuella Brennan. St. Louis, MO: B. Herde.

Aquinas, Thomas. 1952. *Summa Theologica.* Trans. Fathers of the English Dominican Province, rev. Daniel J. Sullivan. Chicago: Encyclopaedia Brittanica.

Arbib, Michael A. 2003. *The Handbook of Brain Theory and Neural Networks.* Cambridge, MA: MIT Press.

Aristotle. 1984. *The Complete Works of Aristotle.* Ed. Jonathan Barnes. Princeton, NJ: Princeton University Press.

Arom, Simha. 1991. *African Polyphony and Polyrhythm: Musical Structure and Methodology.* Cambridge: Cambridge University Press.

Atkinson, Charles M. 1999. Martianus Capella 935 and its Carolingian commentaries. *Journal of Musicology* 17:498–519.

Atkinson, Russell E. 2010. Some thoughts on music pedagogy in the Carolingian era. In *Music Education in the Middle Ages and the Renaissance,* ed. Cynthia J. Cyrus, Susan Forscher Weiss, and Russell E. Murray, 37–51. Bloomington: Indiana University Press.

Augustine of Hippo. 1990. *The Works of Saint Augustine.* Trans. and ed. Edmund Hill and John E. Rotelle. Brooklyn, NY: New City Press.

Augustine of Hippo. 1995. *Against Academicians and The Teacher.* Trans. Peter King. Indianapolis, IN: Hackett.

Augustine of Hippo. 1997. *Œuvres de Saint Augustin.* Trans. and ed. M. Mellet and T. Camelot. Montreal: Institut d'études Augustiniennes.

Augustine of Hippo. 2004. *Expositions of the Psalms 121–150.* Trans. Maria Boulding. Hyde Park, NY: New City Press.

Augustine of Hippo. 2007. *Confessions.* Trans. F. J. Sheed. 2nd ed. Indianapolis, IN: Hackett.

Babbitt, Susan M. 1985. Oresme's *Livre De Politiques* and the France of Charles V. *Transactions of the American Philosophical Society* 75:1–158.

Bach, Carl Philipp Emanuel. 1949. *Essay on the True Art of Playing Keyboard Instruments*. Trans. W. J. Mitchell. New York: W. W. Norton.

Bach, Johann Sebastian. 1995. *J. S. Bach's Precepts and Principles for Playing the Thorough-Bass or Accompanying in Four Parts*. Trans. P. L. Poulin. Oxford: Oxford University Press.

Bacon, Francis. 2000. *The Essayes or Counsels, Civill and Morall*. Ed. Michael Kiernan. Oxford: Clarendon Press.

Bakhtin, Mikhail. 1984. *Problems of Dostoevsky's Poetics*. Trans. and ed. Caryl Emerson. Minneapolis: University of Minnesota Press.

Banich, Marie T., Gregory C. Burgess, Brendan E. Depue, Luka Ruzic, L. Cinnamon Bidwell, Sena Hitt-Laustsen, Yiping P. Du, and Erik G. Willcutt. 2009. The neural basis of sustained and transient attentional control in young adults with ADHD. *Neuropsychologia* 47:3095–3104.

Barker, Andrew. 1990. *Heterophonia* and *Poikilia*: Accompaniments to Greek melody. In *Mousike: Metrica ritmica e musica Greca in memoria di Giovanni Comotti*, ed. Bruno Gentili and Franca Perusino, 41–60. Pisa: Istituti Editoriali e Poligrafici Internazionali.

Bede. 2010. *On the Nature of Things and On Time*. Trans. Calvin B. Kendall and Faith Wallis. Liverpool: Liverpool University Press.

Bent, Ian. 2002. Steps to Parnassus: Contrapuntal theory in 1725 precursors and successors. In *The Cambridge History of Western Music Theory*, ed. Thomas Christensen, 554–602. Cambridge: Cambridge University Press.

Berger, Anna Maria Busse. 2005. *Medieval Music and the Art of Memory*. Berkeley: University of California Press.

Berger, Hans. 1969. *Hans Berger on the Electroencephalogram of Man: The Fourteen Original Reports on the Human Electroencephalogram*. Trans. P. Gloor. Amsterdam: Elsevier.

Bernstein, David W. 2014. John Cage's *Cartridge Music* (1960): "A galaxy reconfigured." *Contemporary Music Review* 33:556–569.

Bett, Henry. 1964. *Johannes Scotus Erigena: A Study in Medieval Philosophy*. New York: Russell & Russell.

Binski, Paul. 2010. Reflections on the "wonderful height and size" of Gothic great churches and the medieval sublime. In *Magnificence and the Sublime in Medieval Aesthetics: Art, Architecture, Literature, Music*, ed. C. Stephen Jaeger, 129–156. New York: Palgrave Macmillan.

Birke, Joachim. 1966. *Christian Wolffs Metaphysik und die zeitgenössische Literatur- und Musiktheorie: Gottsched, Scheibe, Mizler*. Berlin: de Gruyter.

Blackburn, Bonnie J. 2000. Masses based on popular songs and on solmization syllables. In *The Josquin Companion*, ed. Richard Sherr, 1:51–88. Oxford: Oxford University Press.

Boden, Margaret A. 2006. *Mind as Machine: A History of Cognitive Science*. Oxford: Clarendon Press.

Boethius. 1936. *The Theological Tractates*. Trans. and ed. H. F. Stewart and Edward Kennard Rand. Cambridge, MA: Harvard University Press.

Boethius. 1989. *Fundamentals of Music*. Trans. Calvin M. Bower and Claude V. Palisca. New Haven, CT: Yale University Press.

Bonds, Mark Evan. 1991. *Wordless Rhetoric: Musical Form and the Metaphor of the Oration*. Cambridge, MA: Harvard University Press.

Borkowski, J. A. 1997. *Textbook on Roman Law*. 2nd ed. London: Blackstone Press.

Bourdieu, Pierre. 1986. *Distinction: A Social Critique of the Judgement of Taste*. London: Routledge & Kegan Paul.

Bourdieu, Pierre, and Sayad Abdelmalek. 1964. *Le déracinement: La crise de l'agriculture traditionnelle en Algérie*. Paris: Éditions de Minuit.

Bower, Calvin M. 1971. Natural and artificial music: The origins and development of an aesthetic concept. *Musica disciplina* 25:17–33.

Bower, Calvin M. 2002. "Adhuc ex parte et in enigmate cernimus ...": Reflections on the closing chapters of *Musica enchiriadis*. In *Music in the Mirror: Reflections on the History of Music Theory and Literature for the 21st Century*, ed. Andrea Giger and Thomas J. Mathiesen. Lincoln: University of Nebraska Press.

Brinkmann, Reinhold. 1999. The compressed symphony: On the historical content of Schoenberg's Op. 9. In *Schoenberg and His World*, ed. Walter Frisch, 141–161. Princeton, NJ: Princeton University Press.

Brown, E. M. 1983. Neurology and spiritualism in the 1870s. *Bulletin of the History of Medicine* 57:563–577.

Brown, Peter. 2000. *Augustine of Hippo: A Biography*. New ed. Berkeley: University of California Press.

Brown, Thomas. 2006. *Attention Deficit Disorder: The Unfocused Mind in Children and Adults*. New ed. New Haven, CT: Yale University Press.

Bukofzer, Manfred. 1940. Popular polyphony in the Middle Ages. *Musical Quarterly* 26:31–49.

Burbach, Hermann-Josef. 1966. *Studien zur Musikanschauung des Thomas von Aquin*. Regensburg: Bosse.

Burney, Charles. 1957. *A General History of Music: From the Earliest Ages to the Present Period*. Ed. Frank Mercer. New York: Dover.

Butler, Gregory G. 1977. Fugue and rhetoric. *Journal of Music Theory* 21:49–109.

Butt, John. 1997. "A mind unconscious that it is calculating"?: Bach and the rationalist philosophy of Wolff, Leibniz and Spinoza. In *The Cambridge Companion to Bach*, ed. John Butt, 60–71. Cambridge: Cambridge University Press.

Buzsáki, György. 2006. *Rhythms of the Brain*. Oxford: Oxford University Press.

Buzsáki, György. 2010. Neural syntax: Cell assemblies, synapsembles, and readers. *Neuron* 68 (3): 362–385.

Buzsáki, György, Nikos Logothetis, and Wolf Singer. 2013. Scaling brain size, keeping timing: Evolutionary preservation of brain rhythms. *Neuron* 80 (3): 751–764.

Buzsáki Lab. 2016. Accessed October 3. http://www.buzsakilab.com/.

Byrd, Joseph. 1993. *Variations IV*. In *Writings about John Cage*, ed. Richard Kostelanetz, 134–135. Ann Arbor: University of Michigan Press.

Cage, John. 1969. *A Year from Monday: New Lectures and Writings*. Middletown, CT: Wesleyan University Press.

Cage, John. 2011. *Silence: Lectures and Writings*. Middletown, CT: Wesleyan University Press.

Callendar, Clifton. 2016. Performing the irrational: Paul Usher's arrangement of Nancarrow's Study no. 33, Canon 2: $\sqrt{2}$. Accessed September 25. http://conlonnancarrow.org/symposium/papers/callender/irrational.html.

Calvin, William H. 1990. *The Cerebral Symphony: Seashore Reflections on the Structure of Consciousness*. New York: Bantam Books.

Canales, Jimena. 2015. *The Physicist and the Philosopher: Einstein, Bergson, and the Debate That Changed Our Understanding of Time*. Princeton, NJ: Princeton University Press.

Canning, Nancy. 1989. Glenn Gould's contrapuntal radio: Bach to the future. *Ear* 14 (5): 18–23.

Carabine, Deirdre. 2000. *John Scottus Eriugena*. New York: Oxford University Press.

Cardiff, Janet, and George Bures Miller. 2015. The Forty Part Motet (2001). Accessed October 15. http://www.cardiffmiller.com/artworks/inst/motet.html.

Carpenter, Nan Cooke. 1972. *Music in the Medieval and Renaissance Universities*. New York: Da Capo Press.

Carruthers, Mary. 2008. *The Book of Memory*. 2nd ed. Cambridge: Cambridge University Press.

Caspar, Max. 1993. *Kepler*. Trans. C. D. Hellman. New York: Dover.

Castiglione, Baldassarre. 2002. *The Book of the Courtier*. Trans. C. S. Singleton. New York: W. W. Norton.

Caton, Richard. 1875. The electric currents of the brain. *British Medical Journal* 2 (765): 278.

Changeux, Jean-Pierre. 2004. *The Physiology of Truth: Neuroscience and Human Knowledge*. Cambridge, MA: Harvard University Press.

Chernoff, John Miller. 1981. *African Rhythm and African Sensibility: Aesthetics and Social Action in African Musical Idioms*. Chicago: University of Chicago Press.

Cherry, C. 1953. Some experiments on the reception of speech with one and with two ears. *Journal of the Acoustical Society of America* 25:975–979.

Christensen, Thomas S., ed. 2002. *The Cambridge History of Western Music Theory*. Cambridge: Cambridge University Press.

Ciabattoni, Francesco. 2010. *Dante's Journey to Polyphony*. Toronto: University of Toronto Press.

Cicero, Marcus Tullius. 1977. *De re publica; De legibus*. Trans. C. W. Keyes. Cambridge, MA: Harvard University Press.

Clark, Suzannah, and Alexander Rehding, eds. 2001. *Music Theory and Natural Order from the Renaissance to the Early Twentieth Century*. Cambridge: Cambridge University Press.

Clendinning, Jane Piper. 1989. Contrapuntal techniques in the music of György Ligeti. PhD diss., Yale University.

Cohen, David E. 1993. Metaphysics, ideology, discipline: Consonance, dissonance, and the foundations of Western polyphony. *Theoria* 7:1–85.

Comotti, Giovanni. 1989. *Music in Greek and Roman Culture*. Trans. R. V. Munson. Baltimore, MD: Johns Hopkins University Press.

Cornish, Alison. 2000. *Reading Dante's Stars.* New Haven, CT: Yale University Press.

Corwin, Lucille. 2008. "Le Istitutioni Harmoniche" of Gioseffo Zarlino, Part 1: A translation with introduction. PhD diss., City University of New York.

Couperin, François. 1717. *L'art de toucher le clavecin*. Paris.

Crick, Francis. 1989. The recent excitement about neural networks. *Nature* 337 (6203): 129–132.

Crick, Francis. 1994. *The Astonishing Hypothesis: The Scientific Search for the Soul*. New York: Scribner.

Crick, Francis, and Christof Koch. 1990. Towards a neurobiological theory of consciousness. *Seminars in Neuroscience* 2:263–275.

Crighton-Browne, J. 1907. Dexterity and the bend sinister. *Proceedings of the Royal Institution* 18:623–652.

Crocker, Richard. 2011. No polyphony before A.D. 900! In *Strings and Threads: A Celebration of the Work of Anne Draffkorn Kilmer*, ed. Wolfgang Heimpel and Gabriella Frantz-Szabó, 45–58. Winona Lake, IN: Eisenbrauns.

Crowe, Peter. 1981. Polyphony in Vanuatu. *Ethnomusicology* 25:419–432.

Csapo, Eric. 2004. The politics of the new music. In *Music and the Muses: The Culture of Mousike in the Classical Athenian City*, ed. Penelope Murray and Peter Wilson, 207–248. Oxford: Oxford University Press.

Dahlhaus, Carl. 1987. *Schoenberg and the New Music*. Cambridge: Cambridge University Press.

Daly, Ann. 2010. *Done into Dance: Isadora Duncan in America*. Middletown, CT: Wesleyan University Press.

Damasio, Antonio. 2000. *The Feeling of What Happens: Body and Emotion in the Making of Consciousness*. New York: Mariner Books.

Dante Alighieri. 1975. *The Divine Comedy*. Trans. Charles S. Singleton. Princeton, NJ: Princeton University Press.

D'Ardenne, Kimberlee. 2013. The brain is an orchestra. *Psychology Today*, Nov. 15. https://www.psychologytoday.com/blog/quilted-science/201311/the-brain-is-orchestra.

Davidson, Peter. 2005. *The Idea of North*. London: Reaktion Books.

Dean, Jeffrey. 1997. Listening to sacred polyphony c. 1500. *Early Music* 25:611–637.

Dehaene, Stanislas, and Jean-Pierre Changeux. 2011. Experimental and theoretical approaches to conscious processing. *Neuron* 70 (2): 200–227.

Dennett, Daniel C. 1991. *Consciousness Explained*. Boston: Little, Brown.

Descartes, René. 1970. *Philosophical Letters*. Ed. Anthony Kenney. Oxford: Clarendon Press.

Descartes, René. 1996. *Oeuvres de Descartes*. Ed. Charles Adam and Paul Tannery. Paris: Vrin.

Dickinson, P. S., and F. Nagy. 1983. Control of a central pattern generator by an identified modulatory interneurone in Crustacea: II. Induction and modification of plateau properties in pyloric neurones. *Journal of Experimental Biology* 105 (1): 59–82.

Dijk, Piter van. 2002. Aspects of fingering and hand division in Lynar A1. In *Sweelinck Studies*, ed. P. Dirksen, 127–144. Utrecht: STIMU.

Dirksen, Pieter. 2007. *Heinrich Scheidemann's Keyboard Music: Transmission, Style and Chronology*. Aldershot, Hampshire: Ashcroft.

Dolan, Emily I. 2013. *The Orchestral Revolution: Haydn and the Technologies of Timbre*. Cambridge: Cambridge University Press.

Duchesneau, Louise, and Wolfgang Marx, eds. 2011. *György Ligeti: Of Foreign Lands and Strange Sounds*. Woodbridge, Suffolk: Boydell Press.

Duchez, Marie-Elisabeth. 1980. Jean Scot Erigène: Premier lecteur du *De institutione musica* de Boèce? In *Eriugena: Studien zu seinen Quellen*, ed. Werner Beierwaltes. Heidelberg: Carl Winter Verlag.

Duff, P. W. 1971. *Personality in Roman Private Law*. New York: A. M. Kelley.

Duffin, Ross. 2015. Leonardo's lira. *Cleveland Arts* 55 (3): 10–12.

Durkheim, Emile. 2014. *The Division of Labor in Society*. New York: Free Press.

Dwyer, Jim. 2013. Moved to tears at the Cloisters by a ghostly tapestry of music. *New York Times*, Sept. 19.

Eagleman, David. 2015. *The Brain: The Story of You*. New York: Pantheon.

Early Music Consort of London. 2002. *Music of the Gothic Era*. Audio CD. Archiv.

Edler, Arnfried, and Friedrich Wilhelm Riedel, eds. 1996. *Johann Joseph Fux und seine Zeit: Kultur, Kunst und Musik im Spätbarock*. Laaber: Laaber Verlag.

Eling, Paul, Stanley Finger, and Harry Whitaker. 2015. Franz Joseph Gall and music: The faculty and the bump. In *Music, Neurology, and Neuroscience: Historical Connections and Perspectives*, ed. Eckart Altenmüller, Stanley Finger and François Boller, 3–32. Amsterdam: Elsevier.

Ellenberger, Henri F. 1970. *The Discovery of the Unconscious: The History and Evolution of Dynamic Psychiatry*. New York: Basic Books.

Erickson, Raymond. 1992. Eriugena, Boethius and the Neoplatonism of *Musica* and *Scolica enchiriadis*. In *Musical Humanism and Its Legacy: Essays in Honor of Claude V. Palisca*, ed. Nancy Kovaleff Baker and Barbara Russano Hanning. Stuyvesant, NY: Pendragon Press.

Erickson, Raymond, and Claude V. Palisca. 1995. *Musica enchiriadis and Scolica enchiriadis*. New Haven, CT: Yale University Press.

Eriugena, Joahnnes Scottus. 1970. *Iohannis Scotti Annotationes in Marcianum*. Ed. Cora E. Lutz. New York: Kraus Reprint.

Eriugena, Johannes Scottus. 1987. *Periphyseon (The Division of Nature)*. Trans. I. P. Sheldon-Williams, rev. John J. O'Meara. Washington, DC: Dumbarton Oaks.

Eriugena, John Scotus. 1990. *The Voice of the Eagle: Homily on the Prologue to the Gospel of St. John*. Trans. Christopher Bamford. Hudson, NY: Lindisfarne Press.

Evagrius Ponticus. 1970. *The Praktikos: Chapters on Prayer*. Trans. John Eudes Bamberger. Spencer, MA: Cistercian Publications.

Evagrius Ponticus. 2003. *Evagrius of Pontus: The Greek Ascetic Corpus*. Trans. R. E. Sinkewicz. New York: Oxford University Press.

Fassler, Margot E. 1987. Accent, meter, and rhythm in medieval treatises "De Rithmis." *Journal of Musicology* 5:164–190.

Federhofer, Hellmut. 1996. L. Chr. Mizlers Kommentare zu den beiden Büchern des Gradus ad Parnassum von J. J. Fux. In *Johann Joseph Fux und seine Zeit: Kultur, Kunst und Musik im Spätbarock*, ed. Arnfried Edler and Friedrich Wilhelm Riedel, 121–136. Laaber: Laaber Verlag.

Felbick, Lutz. 2012. *Lorenz Christoph Mizler de Kolof: Schüler Bachs und pythagoreischer Apostel der Wolf-fischen Philosophie*. Hildesheim: Olms.

Ferguson, Howard. 1975. *Keyboard Interpretation from the 14th to the 19th Century: An Introduction*. New York: Oxford University Press.

Festinger, Leon. 1956. *When Prophecy Fails*. Minneapolis, MN: University of Minnesota Press.

Festinger, Leon. 1962. *A Theory of Cognitive Dissonance*. Stanford, CA: Stanford University Press.

Festinger, Leon. 1964. *Conflict, Decision, and Dissonance*. Stanford, CA: Stanford University Press.

Finger, Stanley. 2000. *Minds behind the Brain: A History of the Pioneers and Their Discoveries*. Oxford: Oxford University Press.

Frégnac, Yves, Pedro Carelli, Marc Pananceau, and Cyril Monier. 2010. Stimulus-driven coordination of cortical cell assemblies and propagation of gestalt belief in V1. In *Dynamic Coordination in the Brain: From Neurons to Mind*, ed. Christoph von der Malsburg, William A. Phillips and Wolf Singer, 169–192. Cambridge, MA: MIT Press.

Freud, Sigmund. 1963. *Character and Culture*. New York: Collier Books.

Fries, Pascal. 2006. The powers of rhythm. *Science* 314 (5796): 58.

Fuller, Sarah. 1990. Early polyphony. In *The Early Middle Ages to 1300*, ed. Richard Crocker and David Hiley, 485–556. Oxford: Oxford University Press.

Fuller, Sarah. 2002. Organum—discantus—contrapunctus in the Middle Ages. In *The Cambridge History of Western Music Theory*, ed. Thomas Christensen, 477–502. Cambridge: Cambridge University Press.

Fux, Johann Joseph. 1725. *Gradus ad Parnassum*. Vienna: Typis Joannis Petri van Ghelen.

Fux, Johann Joseph. 1742. *Gradus ad Parnassum oder Anführung zur regelmäßigen musikalischen Composition ...* Trans. L. Mizler. Leipzig: Mizler.

Fux, Johann Joseph. 1965. *Study of Counterpoint: From Johann Joseph Fux's* Gradus ad Parnassum. Trans. A. Mann. New York: W. W. Norton.

Gaffurius, Franchinus. 1969. *The Practica Musicae of Franchinus Gafurius*. Trans. I. Young. Madison, WI: University of Wisconsin Press.

Gaffurius, Franchinus. 1993. *The Theory of Music*. New Haven, CT: Yale University Press.

Galilei, Galileo. 2000. *Two New Sciences, Including Centers of Gravity and Force of Percussion*. Trans. S. Drake. Toronto: Wall & Emerson.

Galilei, Vincenzo. 2003. *Dialogue on Ancient and Modern Music*. Trans. Claude V. Palisca. New Haven, CT: Yale University Press.

Gann, Kyle. 1995. *The Music of Conlon Nancarrow*. Cambridge: Cambridge University Press.

Gann, Kyle. 2010. *No Such Thing as Silence: John Cage's 4′33″*. New Haven, CT: Yale University Press.

Gazzaley, Adam, and Larry D. Rosen. 2016. *The Distracted Mind: Ancient Brains in a High-Tech World*. Cambridge, MA: MIT Press.

Gentili, Bruno. 1990. *Poetry and Its Public in Ancient Greece*. Trans. A. Thomas Cole. Baltimore, MD: The Johns Hopkins University Press.

Georgiades, Thrasybulos G. 1973. *Greek Music, Verse, and Dance*. New York: Da Capo Press.

Gerson, Lloyd P. 2003. *Knowing Persons: A Study in Plato*. Oxford: Oxford University Press.

Gevaert, François-Auguste. 1875. *Histoire et théorie de la musique de l'antiquité*. Ghent: C. Annoot-Braeckman.

Gibbon, Edward. 1994. *The History of the Decline and Fall of the Roman Empire*. New York: Penguin.

Gingerich, Owen. 1993. *The Eye of Heaven: Ptolemy, Copernicus, Kepler*. New York: American Institute of Physics.

Girardeau, Gabrielle, Karim Benchenane, Sidney I. Wiener, György Buzsáki, and Michaël B. Zugaro. 2009. Selective suppression of hippocampal ripples impairs spatial memory. *Nature Neuroscience* 12 (10): 1222–1223.

Girardeau, Gabrielle, and Michaël Zugaro. 2011. Hippocampal ripples and memory consolidation. *Current Opinion in Neurobiology* 21 (3): 452–459.

Glymour, Clark. 1991. Freud's androids. In *The Cambridge Companion to Freud*, ed. Jerome Neu, 44–85. Cambridge: Cambridge University Press.

Godman, Stanley. 1957. Bach's copies of Ammerbach's "Orgel oder Instrument Tabulatur" (1571). *Music & Letters* 38:21–27.

Goel, A., and D. V. Buonomano. 2014. Timing as an intrinsic property of neural networks: Evidence from in vivo and in vitro experiments. *Philosophical Transactions of the Royal Society of London, Series B, Biological Sciences* 369 (1637): 20120460.

Goethe, Johann Wolfgang von. 2008. *Faust, Part One*. Trans. David Luke. Oxford: Oxford University Press.

Goffman, Erving. 1974. *Frame Analysis: An Essay on the Organization of Experience*. Cambridge, MA: Harvard University Press.

Goldstein, Jan. 2005. *The Post-Revolutionary Self: Politics and Psyche in France, 1750–1850*. Cambridge, MA: Harvard University Press.

Gould, Glenn. 1984. *The Glenn Gould Reader*. Ed. Tim Page. New York: Vintage.

Grace, Sherrill E. 2015. *Canada and the Idea of North*. Montreal: McGill-Queen's Press.

Grafton, Anthony. 2000. *Leon Battista Alberti: Master Builder of the Italian Renaissance*. New York: Hill and Wang.

Grassmann, Bernd, and Ulrich Weiss. 1988. Unterrichtsplanung im Gespräch: John Cages Radio music, die Auflösung des traditionellen Kunstwerkbegriffs. *Musik & Bildung* 20 (1): 20–23.

Gray, Charles M., Peter König, Andreas K. Engel, and Wolf Singer. 1989. Oscillatory responses in cat visual cortex exhibit inter-columnar synchronization which reflects global stimulus properties. *Nature* 338 (6213): 334–337.

Gresser, Clemens. 1998. "… A music made by everyone": Eine Analytische Annäherung an John Cages Number Pieces. *MusikTexte: Zeitschrift für neue Musik*, no. 76–77: 41–47.

Grinvald, Amiram, Amos Arieli, Misha Tsodyks, and Tal Kenet. 2003. Neuronal assemblies: Single cortical neurons are obedient members of a huge orchestra. *Biopolymers* 68 (3): 422–436.

Haar, James. 1960. Musica mundana: Variations on a Pythagorean theme. PhD diss., Harvard University.

Haar, James. 1976. Some remarks on the "Missa La sol fa re mi." In *Josquin des Prez: Proceedings of the International Josquin Festival Conference*, ed. Edward E. Lowinsky, 564–588. Oxford: Oxford University Press.

Hardcastle, Valerie Gray. 1996. The binding problem and neurobiological oscillations. In *Toward a Science of Consciousness: The First Tucson Discussions and Debates*, ed. Stuart R. Hameroff, Alfred W. Kaszniak and Alwyn C. Scott, 51–65. Cambridge, MA: MIT Press.

Harrington, Anne. 1987. *Medicine, Mind, and the Double Brain: A Study in Nineteenth Century Thought*. Princeton, NJ: Princeton University Press.

Harrington, Anne. 1996. *Reenchanted Science: Holism in German Culture from Wilhelm II to Hitler*. Princeton, NJ: Princeton University Press.

Harris, Max. 2014. *Sacred Folly: A New History of the Feast of Fools*. Ithaca, NY: Cornell University Press.

Hayburn, Robert F. 1979. *Papal Legislation on Sacred Music, 95 A.D. to 1977 A.D.* Collegeville, MN: Liturgical Press.

Hebb, Donald O. 1949. *The Organization of Behavior: A Neuropsychological Theory*. New York: Wiley.

Heller-Roazen, Daniel. 2009. *The Inner Touch. Archaeology of a Sensation*. Repr. ed. New York: Zone Books.

Heller-Roazen, Daniel. 2013. *Dark Tongues: The Art of Rogues and Riddlers*. New York: Zone Books.

Helmholtz, Hermann von. 1954. *On the Sensations of Tone as a Physiological Basis for the Theory of Music*. Trans. and ed. Alexander John Ellis. New York: Dover.

Henry, Paul. 1960. *Saint Augustine on Personality*. New York: Macmillan.

Hermans, Hubert J. M. 1999. The polyphony of the mind: A multi-voiced and dialogical self. In *The Plural Self: Multiplicity in Everyday Life*, ed. John Rowan and Mick Cooper, 107–131. London: Sage.

Herron, Jeannine, ed. 2012. *Neuropsychology of Left-Handedness*. Elsevier.

Hilliard Ensemble. 2000. *Perotin*. Audio CD. ECM.

Himebauch, Daniel. 2016. *Official World's Fastest Guitar Player 2000 BPM*. Accessed Oct. 24. https://www.youtube.com/watch?annotation_id=annotation_3002571639&feature=iv&src_vid=VHIxxH7_g9U&v=CSaRWt4cxao#t=8m30s.

Holbrook, Amy Kusian. 1983. The concept of musical consonance in Greek antiquity and its application in the earliest medieval descriptions of polyphony. PhD diss., University of Washington.

Holton, Gerald. 1986. *The Advancement of Science, and Its Burdens*. Cambridge: Cambridge University Press.

Holton, Gerald. 1988. *Thematic Origins of Scientific Thought: Kepler to Einstein*. Rev. ed. Cambridge, MA: Harvard University Press.

Hoppin, Richard H. 1978. *Medieval Music*. New York: W. W. Norton.

Horn, Stacy. 2013. Encircled by sound: Janet Cardiff's "Forty Part Motet." *Chorus America*, Nov. 21. https://www.chorusamerica.org/singers/encircled-sound-janet-cardiff%E2%80%99s-%E2%80%9Cforty -part-motet%E2%80%9D.

Hucbald of Saint Amand, Guido d'Arezzo, and Johannes Afflighemensis. 1978. *Hucbald, Guido, and John on Music: Three Medieval Treatises*. Ed. Claude V. Palisca, trans. Warren Babb. New Haven, CT: Yale University Press.

Huglo, Michel. 1988. Bilbiographie des éditions et études relatives à la théorie musicale du Moyen Âge. *Acta musicologica* 40:261.

Huglo, Michel. 1993. Organum décrit, organum prescript, organum proscrit, organum écrit. In *Polyphonies de tradition orale: Histoire et traditions vivantes: Actes du colloque de Royamont 1990*, ed. Michel Huglo, Marcel Pérès, and Christian Meyer, 13–22. Paris: Créaphis.

Hui, Alexandra. 2013. *The Psychophysical Ear: Musical Experiments, Experimental Sounds, 1840–1910*. Cambridge, MA: MIT Press.

Hunchuk, Allan Murray, and Fabio B. Dasilva. 1989. Musical spaces: A phenomenological analysis of John Cage's *Variations IV*. *Music Review* 50:281–296.

Hunchuk, Allan Murray, and Fabio B. Dasilva. 1990. Everywhere and nowhere: On John Cage and Maurice Merleau-Ponty. *Music Review* 51 (4): 307–323.

Huot, Sylvia. 1997. *Allegorical Play in the Old French Motet: The Sacred and the Profane in Thirteenth-Century Polyphony*. Stanford, CA: Stanford University Press.

Hüschen, Heinrich. 1970. Albertus Magnus und seine Musikanschauung. In *Speculum Musicae Artis: Festgabe für Heinrich Husmann: Dargebracht von seinen Freunden und Schülern*, ed. Heinz Becker and Reinhard Gerlach. Munich: Wilhelm Fink Verlag.

Husmann, Heinrich. 1966. The practice of organum in the liturgical singing of the Syrian churches of the Near and Middle East. In *Aspects of Medieval and Renaissance Music: A Birthday Offering to Gustav Reese*, ed. J. LaRue. New York: W. W. Norton.

Husserl, Edmund. 1970. *Philosophie première, 1923–24*. Paris: Presses universitaires de France.

Jackson, John. 1905. *Ambidexterity, Or, Two-Handedness and Two-Brainedness*. London: Kegan Paul, Trench, Trübner & Company.

Jackson, John Hughlings. 1866. Clinical remarks on emotional and intellectual language in some cases of disease of the nervous system. *Lancet* 87:174–176.

Jackson, Maggie. 2008. *Distracted: The Erosion of Attention and the Coming Dark Age*. Amherst, NY: Prometheus Books.

James, William. 1950. *The Principles of Psychology*. New York: Dover.

Janata, Petr, Barbara Tillmann, and Jamshed J. Bharucha. 2002. Listening to polyphonic music recruits domain-general attention and working memory circuits. *Cognitive, Affective & Behavioral Neuroscience* 2:121–140.

Janet Cardiff: Forty-Part Motet. 2015. National Gallery of Canada. Accessed Oct. 16. https://www.gallery.ca/en/ see/exhibitions/current/details/janet-cardiff-forty-part-motet-6857.

Janet Cardiff's 40 Part Motet. 2015. Accessed Oct. 15. http://www.studio360.org/story/janet-cardiffs-40-part -motet/.

Jankélévitch, Vladimir. 2003. *Music and the Ineffable*. Trans. C. Abbate. Princeton, NJ: Princeton University Press.

Jeauneau, Édouard. 1978. *Quatre thèmes Érigéniens*. Montreal: Institut d'études médiévales Albert-le-Grand.

Jenkins, Todd S., ed. 2004. *Free Jazz and Free Improvisation: An Encyclopedia*. Westport, CT: Greenwood Press.

Jeppesen, Knud. 2012. *The Style of Palestrina and the Dissonance*. New York: Dover.

Jeppesen, Knud. 2013. *Counterpoint: The Polyphonic Vocal Style of the Sixteenth Century*. New York: Dover.

Johnson, Calvert, ed. 1994. *Historical Organ Techniques and Repertoire: An Historical Survey of Organ Performance Practices and Repertoire*. Boston: Wayne Leupold Editions.

Johnson, Julene K., and Amy B. Graziano. 2015. Some early cases of aphasia and the capacity to sing. In *Music, Neurology, and Neuroscience: Historical Connections and Perspectives*, ed. Eckart Altenmüller, Stanley Finger, and François Boller, 73–89. Amsterdam: Elsevier.

Johnson, Julian. 1999. *Webern and the Transformation of Nature*. Cambridge: Cambridge University Press.

Jordania, Joseph. 2006. *Who Asked the First Question? The Origins of Human Choral Singing, Intelligence, Language and Speech*. Tbilisi: Logos.

Jordania, Joseph. 2015. *Choral Singing in Human Culture and Evolution*. Saarbrücken: Lambert Academic Publishing.

Jost, Ekkehard. 1994. *Free Jazz*. New York: Da Capo Press.

Jeauneau, Édouard. 1978. *Quatre thèmes Érigéniens*. Montreal: Institut d'études médiévales Albert-le-Grand.

Kahn, Douglas. 1997. John Cage: Silence and silencing. *Musical Quarterly* 81:556–598.

Kania, Andrew. 2010. Silent music. *Journal of Aesthetics and Art Criticism* 68 (4): 343–353.

Kant, Immanuel. 1965. *Critique of Pure Reason*. Trans. N. K. Smith. New York: St. Martin's Press.

Karmarkar, Uma R., and Dean V. Buonomano. 2007. Timing in the absence of clocks: Encoding time in neural network states. *Neuron* 53 (3): 427–438.

Kelly, Thomas Forrest. 2015. *Capturing Music: The Story of Notation*. New York: W. W. Norton.

Kepler, Johannes. 1997. *The Harmony of the World*. Trans. E. J. Aiton, A. M. Duncan and J. V. Field. Philadelphia: American Philosophical Society.

Kierkegaard, Søren. 1964. *Purity of Heart Is to Will One Thing: Spiritual Preparation for the Office of Confession*. Trans. D. V. Steere. New York: Harper & Row.

Kilmer, Anne Draffkorn. 1974. The cult song with music from Ancient Ugarit: Another interpretation. *Revue d'Assyriologie et d'Archeologie Orientale* 68:69–82.

Kilmer, Anne Draffkorn, Robert R. Brown, and Richard L. Crocker. 1976. *Sounds from Silence: Recent Discoveries in Ancient Near Eastern Music*. Berkeley: Bit Enki Publications.

Kluft, Richard P., and Catherine G. Fine, eds. 1993. *Clinical Perspectives on Multiple Personality Disorder* Washington, DC: American Psychiatric Press.

Knapp, Janet. 1990. Polyphony at Notre Dame of Paris. In *The Early Middle Ages to 1300*, ed. Richard Crocker and David Hiley, 557–635. Oxford: Oxford University Press.

Koch, Christof, and Francis Crick. 2003. Some further ideas regarding the neuronal basis of awareness. In *Large-Scale Neuronal Theories of the Brain*, ed. Christof Koch and Joel L. Davis, 93–109. Cambridge, MA: MIT Press.

Koch, Christof, Marcello Massimini, Melanie Boly, and Giulio Tononi. 2016. Neural correlates of consciousness: Progress and problems. *Nature Reviews Neuroscience* 17 (5): 307–321.

Koelsch, Stefan. 2013. *Brain and Music*. Oxford: Wiley-Blackwell.

Koenig, Gottfried Michael. 1991. *Ästhetische Praxis: Texte zur Musik*. Saarbrücken: Pfau.

Kosovske, Yonit Lea. 2011. *Historical Harpsichord Technique: Developing La douceur du toucher*. Bloomington: Indiana University Press.

Kostelanetz, Richard. 1988. Glenn Gould as a radio composer. *Massachusetts Review* 29:557–570.

Kostelanetz, Richard. 2003. *Conversing with Cage*. 2nd ed. New York: Routledge.

Kursell, Julia. 2013. Experiments on sound color in music and acoustics: Helmholtz, Schoenberg, and Klangfarbenmelodie. *Osiris* 28:191–211.

LaFleur, Karl, Kaitlin Cassady, Alexander Doud, Kaleb Shades, Eitan Rogin, and Bin He. 2013. Quadcopter control in three-dimensional space using a noninvasive motor imagery-based brain–computer interface. *Journal of Neural Engineering* 10 (4): 046003.

Lahire, Bernard. 2011. *The Plural Actor*. Trans. D. Fernbach. Cambridge: Polity Press.

La Mettrie, Julien Offray de. 1912. *Man a Machine*. Trans. Gertrude Carman Bussey and Mary Whiton Calkins. Chicago: Open Court.

Lancaster, Brian. 1999. The multiple brain and the unity of experience. In *The Plural Self: Multiplicity in Everyday Life,* ed. John Rowan and Mick Cooper, 132–150. London: Sage.

Larriva-Sahd, Jorge A. 2014. Some predictions of Rafael Lorente de Nó 80 years later. *Frontiers in Neuroanatomy* 8:147.

Lashley, Karl S. 1963. *Brain Mechanisms and Intelligence: A Quantitative Study of Injuries to the Brain*. New York: Dover.

Leach, Elizabeth Eva. 2009. Music and masculinity in the Middle Ages. In *Masculinity and Western Musical Practice*, ed. Ian D. Biddle and Kirsten Gibson, 22–39. Burlington, VT: Ashgate.

Lee, Albert K., and Matthew A. Wilson. 2002. Memory of sequential experience in the hippocampus during slow wave sleep. *Neuron* 36 (6): 1183–1194.

Leech-Wilkinson, Daniel. 1990. *Machaut's Mass: An Introduction*. Oxford: Clarendon Press.

Leech-Wilkinson, Daniel. 2002. *The Modern Invention of Medieval Music: Scholarship, Ideology, Performance*. Cambridge: Cambridge University Press.

Leitmeir, Christian. 2007. Arguing with spirituality against spirituality: A Cistercian apologia for mensural music by Petrus Dictus Palma Ociosa (1336). *Archa Verbi* 4:115–159.

Leman, Marc. 2008. *Embodied Music Cognition and Mediation Technology*. Cambridge, MA: MIT Press.

Levitin, Daniel J. 2007. *This Is Your Brain on Music: The Science of a Human Obsession*. New York: Plume/Penguin.

Lewis, Jerome. 2013. A cross-cultural perspective on the significance of music and dance to culture and society: Insight from BaYaka Pygmies. In *Language, Music, and the Brain*, ed. Michael A. Arbib, 45–65. Cambridge, MA: MIT Press.

Ligeti, György. 1983. *György Ligeti in Conversation with Péter Várnai, Josef Häusler, Claude Samuel, and Himself*. London: Eulenburg.

Lindley, Mark. 1989. Early fingering: Some editing problems and some new readings for J. S. Bach and John Bull. *Early Music* 17 (1): 60–69.

Lindley, Mark. 1990. Keyboard fingerings and articulation. In *Performance Practice*, ed. Howard Mayer Brown and Stanley Sadie, 2:186–203. New York: W. W. Norton.

Lindley, Mark. 1993. *Ars Ludendi: Early German Keyboard Fingerings, c. 1525–c. 1625*. Neuhof, Germany: Tre Fontane.

Lindley, Mark, and Maria Boxall. 1992. *Early Keyboard Fingerings: A Comprehensive Guide*. Mainz: Schott.

Litweiler, John. 1984. *The Freedom Principle: Jazz after 1958*. New York: W. Morrow.

Little, Meredith, and Natalie Jenne. 2001. *Dance and the Music of J. S. Bach*. Expanded ed. Bloomington: Indiana University Press.

Llinás, Rodolfo R. 1988. The intrinsic electrophysiological properties of mammalian neurons: Insights into central nervous system function. *Science* 242 (4886): 1654.

Llinás, Rodolfo R. 2001. *I of the Vortex: From Neurons to Self*. Cambridge, MA: MIT Press.

Lomax, Alan. 1962. Song structure and social structure. *Ethnology* 1:425–451.

Lomax, Alan. 1968. *Folk Song Style and Culture*. Washington, DC: American Association for the Advancement of Science.

Lorch, Marjorie Perlman, and Samuel H. Greenblatt. 2015. Singing by speechless (aphasic) children: Victorian medical observations. In *Music, Neurology, and Neuroscience: Historical Connections and Perspectives*, ed. Eckart Altenmüller, Stanley Finger and François Boller, 53–72. Amsterdam: Elsevier.

Lorente de Nó, Rafael. 1933. Studies on the structure of the cerebral cortex: I. Area entorhinalis. *Journal für Psychologie und Neurologie* 45:381–438.

Lorente de Nó, Rafael. 1934. Studies on the structure of the cerebral cortex: II. Continuation of the study of the ammonic system. *Journal für Psychologie und Neurologie* 46:113–177.

Lorente de Nó, Rafael. 1951. Cerebral cortex: Architecture, intracortical connections, motor projections. In *Physiology of the Nervous System*, 3rd rev. ed., ed. John F. Fulton, 288–312. New York: Oxford University Press.

Loukopoulos, Loukia D., R. Key Dismukes, Immanuel Barshi. 2009. *The Multitasking Myth: Handling Complexity in Real-World Operations*. Farnham, Surrey: Ashgate.

Lowell, Amy. 1916. *Men, Women and Ghosts*. New York: Macmillan.

Loyola, Ignatius. 1992. *The Autobiography of St. Ignatius Loyola with Related Documents*. Trans. Joseph F. O'Callaghan, ed. John C. Olin. New York: Fordham University Press.

Lucier, Alvin, and Douglas Stone. 2012. *Chambers: Scores by Alvin Lucier*. Middletown, CT: Wesleyan University Press.

Luys, J. B. 1879. Études sur le dédoublement des opérations cérébrales et sur le rôle isolé de chaque hémisphère dans les phénomènes de la pathologie mentale. *Bulletin de l'académie de médicine*, 2nd series, 8: 516–534, 547–565.

Luys, J. B. 1889. Le dédoublement cérébral du pianiste. *Revue de l'hypnotisme et de la psychologie physiologique* 3: 282–284.

Macey, Patrick. 1998. *Bonfire Songs: Savonarola's Musical Legacy*. Oxford: Clarendon Press.

Macey, Patrick. 1999. *Savonarolan Laude, Motets, and Anthems*. Madison, WI: A-R Editions.

MacInnis, John Christian. 2014. "The harmony of all things": Music, soul, and cosmos in the writings of John Scottus Eriugena. PhD diss., Florida State University.

Maconie, Robin. 1990. *The Works of Karlheinz Stockhausen*. Oxford: Clarendon Press.

Maingret, Nicolas, Gabrielle Girardeau, Ralitsa Todorove, Marie Goutierre, and Michaël Zugaro. 2016. Hippocampo-cortical coupling mediates memory consolidation during sleep. *Nature Neuroscience* 19:959–966.

Margulis, Elizabeth Hellmuth. 2014. *On Repeat: How Music Plays the Mind*. New York: Oxford University Press.

Martianus Capella. 1977. *The Marriage of Philology and Mercury*. Trans. W. H. Stahl, R. Johnson and E. L. Burge. New York: Columbia University Press.

Mathiesen, Thomas J. 1999. *Apollo's Lyre: Greek Music and Music Theory in Antiquity and the Middle Ages*. Lincoln: University of Nebraska Press.

Mauer, Barry. 2010. Glenn Gould and the new listener. *Performance Research* 15:103–108.

Mazzeo, Joseph Anthony. 1962. St. Augustine's rhetoric of silence. *Journal of the History of Ideas* 23:175–196.

Mazzola, Guerino, and Paul B. Cherlin. 2009. *Flow, Gesture, and Spaces in Free Jazz: Towards a Theory of Collaboration*. New York: Springer.

McClelland, James L. 1989. Parallel distributed processing: Implications for cognition and development. In *Parallel Distributed Processing: Implications for Psychology and Neurobiology*, ed. R. G. M. Morris. Oxford: Clarendon Press.

McClelland, James L., and David E. Rumelhart, and the PDP Research Group. 1986. *Parallel Distributed Processing: Explorations in the Microstructure of Cognition*. Cambridge, MA: MIT Press.

McGrath, Elizabeth. 2009. Platonic myths in Renaissance iconography. In *Plato's Myths*, ed. Catalin Partenie, 206–238. Cambridge: Cambridge University Press.

McKinney, Timothy R. 2016. *Adrian Willaert and the Theory of Interval Affect: The Musica Nova Madrigals and the Novel Theories of Zarlino and Vicentino*. Farnham, Surrey: Ashgate.

McKinnon, James, ed. 1987. *Music in Early Christian Literature*. Cambridge: Cambridge University Press.

McNeil, Nicole M. 2007. U-shaped development in math: 7-year-olds outperform 9-year-olds on equivalence problems. *Developmental Psychology* 43 (3): 687–695.

McNeill, William Hardy. 1995. *Keeping Together in Time: Dance and Drill in Human History*. Cambridge, MA: Harvard University Press.

McNeilly, Kevin. 1996. Listening, Nordicity, community: Glenn Gould's *The Idea of North*. *Essays on Canadian Writing* 59: 87–104.

Meconi, Honey. 1998. Listening to sacred polyphony. *Early Music* 26:375–379.

Mehta, Mayank. 2007. Fascinating rhythm. *Nature* 446 (7131): 27.

Merce Cunningham: Dance Capsules. 2015. Accessed Dec. 7. http://dancecapsules.mercecunningham.org/overview.cfm?capid=46064.

Mersenne, Marin. 1963. *Harmonie Universelle, contenant la théorie et la pratique de la musique.* Paris: Centre national de la recherche scientifique.

Michelson, Larry K., and William J. Ray, eds. 2013. *Handbook of Dissociation: Theoretical, Empirical, and Clinical Perspectives.* New York: Springer Science & Business Media.

Middlebrook, Diane. 1992. *Anne Sexton: A Biography.* New York: Vintage.

Minsky, Marvin. 1986. *The Society of Mind.* New York: Simon & Schuster.

Mithen, Steven. 2006. *The Singing Neanderthals: The Origins of Music, Language, Mind and Body.* Cambridge, MA: Harvard University Press.

Monson, Craig A. 2002. The Council of Trent revisited. *Journal of the American Musicological Society* 55:1–37.

Morley, Thomas. 1973. *A Plain and Easy Introduction to Practical Music.* 2nd ed., Ed. Alec Harman. New York: W. W. Norton.

Moroney, Davitt. 2003. Couperin, Marpurg and Roesner: A Germanic "Art de toucher le clavecin," or a French "Wahre Art"? In *The Keyboard in Baroque Europe,* ed. Christopher Hogwood, 111–130. Cambridge: Cambridge University Press.

Moroney, Davitt. 2007. Alessandro Striggio's Mass in forty and sixty parts. *Journal of the American Musicological Society* 60:1–70.

Morris, R. G. M., ed. 1989. *Parallel Distributed Processing: Implications for Psychology and Neurobiology.* Oxford: Clarendon Press.

Mullally, Robert. 2011. *The Carole: A Study of a Medieval Dance.* Farnham, Surrey: Ashgate.

Nagy, F., and P. S. Dickinson. 1983. Control of a central pattern generator by an identified modulatory interneurone in Crustacea: I. Modulation of the pyloric motor output. *Journal of Experimental Biology* 105 (1): 33–58.

Nettl, Bruno. 2010. *Nettl's Elephant: On the History of Ethnomusicology.* Urbana: University of Illinois Press.

Nettl, Bruno. 2015. *The Study of Ethnomusicology: Thirty-Three Issues and Concerns.* 3rd ed. Urbana: University of Illinois Press.

Neumann, Anyssa. 2011. Ideas of North: Glenn Gould and the aesthetic of the sublime. *voiceXchange* 5: 35–46.

Nicomachus. 1994. *The Manual of Harmonics of Nicomachus the Pythagorean.* Trans. Flora R. Levin. Grand Rapids, MI: Phanes Press.

Niedermeyer, Ernst, and F. H. Lopes da Silva, eds. 2005. *Electroencephalography: Basic Principles, Clinical Applications, and Related Fields.* Philadelphia, PA: Lippincott Williams & Wilkins.

Niedt, Friederich Erhardt. 1989. *The Musical Guide.* Trans. P. L. Poulin and I. C. Taylor. Oxford: Oxford University Press.

Nietzsche, Friedrich. 1997. *Daybreak: Thoughts on the Prejudices of Morality.* Trans. M. Clark and B. Leiter. Cambridge: Cambridge University Press.

Oresme, Nicole. 1961. *Quaestiones super geometriam Euclidis.* Ed. H. L. L. Busard. Leiden: Brill.

Oresme, Nicole. 1968a. *Le livre du ciel et du monde.* Trans. A. D. Menut. Madison, WI: University of Wisconsin Press.

Oresme, Nicole. 1968b. *Nicole Oresme and the Medieval Geometry of Qualities and Motions: A Treatise on the Uniformity and Difformity of Intensities Known as* Tractatus de configurationibus qualitatum et motuum. Trans. M. Clagett. Madison, WI: University of Wisconsin Press.

Oresme, Nicole. 1971. *Nicole Oresme and the Kinematics of Circular Motion:* Tractatus de commensurabilitate vel incommensurabilitate motuum celi. Trans. E. Grant. Madison, WI: University of Wisconsin Press.

Oresme, Nicole. 1985. *Nicole Oresme and The Marvels of Nature: A Study of His* De causis mirabilium *with Critical Edition, Translation, and Commentary.* Trans. B. Hansen. Toronto: Pontifical Institute of Mediaeval Studies.

Page, Christopher. 1986. *Voices and Instruments of the Middle Ages: Instrumental Practice and Songs in France, 1100–1300*. Berkeley: University of California Press.

Page, Christopher. 1990a. Polyphony before 1400. In *Performance Practice*, ed. Howard Mayer Brown and Stanley Sadie, 1:79–104. New York: W. W. Norton.

Page, Christopher. 1990b. *The Owl and the Nightingale: Musical Life and Ideas in France, 1100–1300*. Berkeley: University of California Press.

Page, Christopher. 1993. *Discarding Images: Reflections on Music and Culture in Medieval France*. Oxford: Clarendon Press.

Page, Christopher. 2007. *Summa Musice: A Thirteenth-Century Manual for Singers*. Cambridge: Cambridge University Press.

Page, Christopher. 2010. *The Christian West and Its Singers: The First Thousand Years*. New Haven, CT: Yale University Press.

Palestrina, Giovanni Pierluigi da. 1975. *Pope Marcellus Mass: An Authoritative Score, Backgrounds and Sources, History and Analysis, Views and Comments*. Ed. Lewis Lockwood. New York: W. W. Norton.

Palisca, Claude V. 1985. *Humanism in Italian Renaissance Musical Thought*. New Haven, CT: Yale University Press.

Panofsky, Erwin. 1957. *Gothic Architecture and Scholasticism*. New York: Meridian Books.

Peacocke, A. R., and Grant Gillett. 1987. *Persons and Personality: A Contemporary Inquiry*. Oxford: Blackwell.

Pederson, O. 1981. The origins of the Theorica Planetarum. *Journal for the History of Astronomy* 12:113–123.

Penfield, Wilder. 1975. *The Mystery of the Mind: A Critical Study of Consciousness and the Human Brain*. Princeton, NJ: Princeton University Press.

Penfield, Wilder, and Herbert H. Jasper. 1954. *Epilepsy and the Functional Anatomy of the Human Brain*. New York: Little, Brown.

Penttonen, Markku, and György Buzsáki. 2003. Natural logarithmic relationship between brain oscillators. *Thalamus Related Systems* 2:145–152.

Peretz, Isabelle, and Robert Zatorre, eds. 2003. *The Cognitive Neuroscience of Music*. Oxford: Oxford University Press.

Perlman, Marc. 2004. *Unplayed Melodies: Javanese Gamelan and the Genesis of Music Theory*. Berkeley: University of California Press.

Pesic, Peter. 2003. *Abel's Proof: An Essay on the Sources and Meaning of Mathematical Unsolvability*. Cambridge, MA: MIT Press.

Pesic, Peter. 2014a. Francis Bacon, violence, and the motion of liberty: The Aristotelian background. *Journal of the History of Ideas* 75:69–90.

Pesic, Peter. 2014b. *Music and the Making of Modern Science*. Cambridge, MA: MIT Press.

Petzold, Charles. 2008. *The Annotated Turing: A Guided Tour through Alan Turing's Historic Paper on Computability and the Turing Machine*. Indianapolis, IN: Wiley.

Pfurtscheller, G., and A. Berghold. 1989. Patterns of cortical activation during planning of voluntary movement. *Electroencephalography and Clinical Neurophysiology* 72 (3): 250–258.

Phillips, Nancy. 1984. *Musica* and *Scolica enchiriadis*: The literary, theoretical, and musical sources. PhD diss., New York University.

Pirrotta, Nino. 1968. Dante musicus: Gothicism, Scholasticism, and music. *Speculum* 43 (2): 245–257.

Plato. 1997. *Complete Works*. Ed. John M. Cooper. Indianapolis, IN: Hackett.

Plomp, R., and W. J. M. Levelt. 1965. Tonal consonance and critical bandwidth. *Journal of the Acoustical Society of America* 38:548–560.

Plotinus. 1992. *The Enneads*. Trans. S. Mackenna. Burdett, NY: Larson Publications.

Plowright, Piers. 2008. Muzak, music, and monologues: The mind of Glenn Gould as revealed in his radio documentaries. *GlennGould* 13:42–45.

Pöhlmann, Egert. 1995. Metrica e ritmica nella poesia e nella musica greca antica. In *Mousike: Metrica ritmica e musica Greca in memoria di Giovanni Comotti*, ed. Bruno Gentili and Franca Perusino, 3–15. Pisa: Istituti editoriali e poligrafici internazionali.

Pöhlmann, Egert, and Martin L. West. 2001. *Documents of Ancient Greek Music: The Extant Melodies and Fragments*. Oxford: Clarendon Press.

Pound, Ezra. 1918. *Pavannes and Divisions*. New York: Alfred A. Knopf.

Pritchett, James. 1996. *The Music of John Cage*. Cambridge: Cambridge University Press.

Proclus. 1963. *The Elements of Theology*. Trans. E. R. Dodds. Oxford: Oxford University Press.

Proust, Marcel. 2003a. *Finding Time Again*. Trans. I. Patterson. London: Penguin Classic.

Proust, Marcel. 2003b. *Swann's Way*. Trans. L. Davis. New York: Penguin Books.

Proust, Marcel. 2005a. *The Guermantes Way*. Trans. M. Treharne. New York: Penguin Classics.

Proust, Marcel. 2005b. *Sodom and Gomorrah*. Trans. J. Sturrock. New York: Penguin Classics.

Radkau, Joachim. 2009. *Max Weber: A Biography*. Cambridge: Polity.

Rámon y Cajal, Santiago. 1954. *Neuron Theory or Reticular Theory? Objective Evidence of the Anatomical Unity of Nerve Cells*. Trans. M. Ubeda Purkiss and Clement A. Fox. Madrid: Consejo Superior de Investigaciones Cientificas.

Rand, E. K. 1940. How much of the *Annotationes in Marcianum* is the work of John the Scot? *Transactions and Proceedings of the American Philological Association* 71:501–523.

Reese, Gustave. 1959. *Music in the Renaissance*. New York: W. W. Norton.

Rehding, Alexander. 2013. Of sirens old and new. In *The Oxford Handbook of Mobile Music*, ed. Sumanth S. Gopinath and Jason Stanyek, 77–106. Oxford: Oxford University Press.

Remigius of Auxerre. 1962. *Remigii Autissiodorensis Commentum in Martianum Capellam*. Ed. Cora E. Lutz. Leiden: E. J. Brill.

Reti, Rudolph. 1978. *Tonality, Atonality, Pantonality: A Study of Some Trends in Twentieth Century Music*. Westport, CT: Greenwood Press.

Revill, David. 1992. *The Roaring Silence: John Cage, a Life*. New York: Arcade Publishing.

Riemann, Hugo. 1974. *History of Music Theory: Polyphonic Theory to the Sixteenth Century*. Trans. R. H. Haggh. New York: Da Capo Press.

Roberts, Alexander, James Donaldson, A. Cleveland Coxe, Allan Menzies, Ernest Cushing Richardson, and Bernhard Pick. 1885. *The Ante-Nicene Fathers: Translations of the Writings of the Fathers Down to A.D. 325*. Buffalo, NY: The Christian Literature Publishing Company.

Robertson, Anne Walters. 2002. *Guillaume de Machaut and Reims: Context and Meaning in His Musical Works*. Cambridge: Cambridge University Press.

Rokseth, Yvonne. 1947. Dances cléricales du XIIIe siècle. In *Mélanges 1945 du publications de la Faculté des Lettres de Strasbourg*, 93–126. Paris: Societé d'édition Les Belles Lettres.

Roth, Herman. 1926. *Elemente der Stimmführung*. Stuttgart: Grüninger.

Rowan, John, and Mick Cooper, eds. 1999. *The Plural Self: Multiplicity in Everyday Life*. London: Sage.

Sachs, Curt. 1962. *The Wellsprings of Music*. The Hague: M. Nijhoff.

Sacks, Oliver. 1995. *An Anthropologist on Mars: Seven Paradoxical Tales*. New York: Knopf.

Sacks, Oliver. 2008. *Musicophilia: Tales of Music and the Brain*. Rev. and exp. ed. New York: Vintage.

Shanzer, Danuta. 1986. *A Philosophical and Literary Commentary on Martianus Capella's* De Nuptiis Philologiae et Mercurii *Book 1*. Berkeley: University of California Press.

Scherer, Reinhold, and Gert Pfurtscheller. 2013. Thought-based interaction with the physical world. *Trends in Cognitive Sciences* 17 (10): 490–492.

Schneider, Marius. 1969. *Geschichte der Mehrstimmigkeit: Historische und phänomenologische Studien*. Tutzing: H. Schneider.

Schrade, Leo, ed. 1960. *La représentation d'Edipo Tiranno au Teatro Olimpico (Vicence, 1585)*. Paris: Éditions du Centre National de la Recherche Scientifique.

Schueller, Herbert M. 1988. *The Idea of Music: An Introduction to Musical Aesthetics in Antiquity and the Middle Ages*. Kalamazoo, MI: Medieval Institute Publications, Western Michigan University.

Shultis, Christopher. 1995. Silencing the sounded self: John Cage and the intentionality of nonintention. *Musical Quarterly* 79:312–350.

Silverman, Kenneth. 2012. *Begin Again: A Biography of John Cage*. Evanston, IL: Northwestern University Press.

Singer, Wolf. 2001. Consciousness and the binding problem. *Annals of the New York Academy of Sciences* 929 (1): 123–146.

Singer, Wolf. 2005. Putative role of oscillations and synchrony in cortical signal processing and attention. In *Neurobiology of Attention*, ed. Laurent Itti, Geraint Rees and John K. Tsotsos, 526–533. Amsterdam: Elsevier.

Smolensky, Paul. 1986. Information processing in dynamical systems: Foundations of harmony theory. In *Parallel Distributed Processing: Explorations in the Microstructure of Cognition*, ed. James L. McClelland, David E. Rumelhart, and PDP Research Group, 1:216–271. Cambridge, MA: MIT Press.

Sperry, Roger. 1968. Hemispheric disconnection and unity in conscious awareness. *American Psychologist* 23:723–733.

Spinoza, Benedictus de. 2000. *Ethics*. Trans. G. H. R. Parkinson. Oxford: Oxford University Press.

Spitta, Philipp. 1979. *Johann Sebastian Bach: His Work and Influence on the Music of Germany, 1685–1750*. Trans. C. Bell and J. A. Fuller-Maitland. Mineola, NY: Dover.

Steege, Benjamin. 2012. *Helmholtz and the Modern Listener*. Cambridge: Cambridge University Press.

Steinitz, Richard. 2003. *György Ligeti: Music of the Imagination*. Boston: Northeastern University Press.

Stevenson, Robert Louis. 1991. *The Strange Case of Dr. Jekyll and Mr. Hyde*. New York: Dover.

Strunk, W. Oliver, and Leo Treitler, eds. 1998. *Source Readings in Music History*. New York: W. W. Norton.

Stryker, Michael P. 1989. Is Grandmother an oscillation? *Nature* 338 (6213): 297–298.

Stumpf, Carl. 2012. *The Origins of Music*. Trans. and ed. David Trippett. Oxford: Oxford University Press.

Sudnow, David. 2001. *Ways of the Hand: A Rewritten Account*. Cambridge, MA: MIT Press.

Sullivan, Blair. 1997. The polyphony of the spheres. *Viator* 28:33–43.

Swafford, Jan. 1997. *Johannes Brahms: A Biography*. New York: Alfred A. Knopf.

Swartz, Barbara E. 1998. The advantages of digital over analog recording techniques. *Electroencephalography and Clinical Neurophysiology* 106:113–117.

Synan, Edward A. 1964. An Augustinian testimony to polyphonic music? *Musica Disciplina* 18:3–6.

Szlezak, Thomas A. 1999. *Reading Plato*. New York: Routledge.

Taruskin, Richard. 2005. *The Oxford History of Western Music*. Oxford: Oxford University Press.

Taylor, Charles. 1989. *Sources of the Self: The Making of the Modern Identity*. Cambridge, MA: Harvard University Press.

Terwen, Jan Willem, and Hilary Staples. 2005. Jan Land en Isaac Groneman als Pioniers in de etnomusicologie. *Tijdschrift van de Koninklijke Vereniging voor Nederlandse Muziekgeschiedenis* 55:163–194.

Teuscher, Christof. 2012. *Turing's Connectionism: An Investigation of Neural Network Architectures*. London: Springer.

Thacker, Andrew. 2004. Unrelated beauty: Amy Lowell, polyphonic prose, and the imagist city. In *Amy Lowell, American Modern*, ed. Adrienne Munich and Melissa Bradshaw, 104–119. New Brunswick, NJ: Rutgers University Press.

Thorndike, Lynn. 1941. Invention of the mechanical clock about 1271 A.D. *Speculum* 16:242–243.

Tinctoris, Johannes. 1961. *The Art of Counterpoint: Liber de arte contrapuncti*. Trans. A. Seay. American Institute of Musicology.

Tomlinson, Gary. 1999. *Metaphysical Song: An Essay on Opera*. Princeton, NJ: Princeton University Press.

Tonus Peregrinus. 2005. *Leonin, Perotin: Sacred Music from Notre-Dame Cathedral*. Audio CD. Naxos.

Torrell, Jean-Pierre, and Robert Royal. 2005. *Saint Thomas Aquinas,* vol. 1: *The Person and His Work*. Rev. ed. Washington, DC: Catholic University of America Press.

Trendelenburg, Adolf. 1910. A contribution to the history of the word person. *Monist* 6:336–363.

Troeger, Richard. 2003. *Playing Bach on the Keyboard: A Practical Guide*. Pompton Plains, NJ: Amadeus Press.

Tulk, Lorne. 2000. Aspects of Glenn Gould: Glenn Gould and the radio documentaries. *Collected Work: Glenn Gould Gathering: Perspectives on the Man and the Musician*. 6 (2): 65–68.

Türk, Daniel Gottlob. 1982. *School of Clavier Playing, Or, Instructions in Playing the Clavier for Teachers and Students*. Trans. R. H. Haggh. Lincoln: University of Nebraska Press.

van der Werf, Hendrik. 1993. *The Oldest Extant Part Music and the Origin of Western Polyphony*. Rochester, NY: H. van der Werf.

van Deusen, Nancy. 1995. *Theology and Music at the Early University: The Case of Robert Grosseteste and Anonymous IV*. Leiden: E. J. Brill.

Venable, Bruce. 1993. The name and nature of the person. In *Essays in Honor of Robert Bart*, ed. Cary Stickney, 260–274. Annapolis, MD: St. John's College Press.

von Neumann, John. 2000. *The Computer and the Brain*. 2nd ed. New Haven, CT: Yale University Press.

Waeltner, Ernst Ludwig. 1977. *Organicum Melos: Zu Musikanschauung des Iohannes Scottus (Eriugena)*. Munich: Verlag der Bayerischen Akademie der Wissenschaften.

Waite, William G. 1973. *The Rhythm of Twelfth-Century Polyphony: Its Theory and Practice*. Westport, CT: Greenwood Press.

Walker, D. P, ed. 1963. *Musique des intermèdes de "La Pellegrina."* Paris: Éditions du Centre National de la Recherche Scientifique.

Walker, Paul. 2004. *Theories of Fugue from the Age of Josquin to the Age of Bach*. Rochester: University Rochester Press.

Warburg, Abby. 1999. Theatrical costumes for the Intermedi of 1589. In *The Renewal of Pagan Antiquity*, trans. David Britt, 349–401; also 495–547 (with translator's addenda). Los Angeles: Getty Research Institute Publications.

Watson, Alan. 1967. *The Law of Persons in the Later Roman Republic*. Oxford: Clarendon Press.

Watson, Brendon O., and György Buzsáki. 2015a. Sleep, memory and brain rhythms. *Daedalus* 144 (1): 67–82.

Watson, Brendon O., and György Buzsáki. 2015b. Neural syntax in mental disorders. *Biological Psychiatry* 77 (12): 998–1000.

Weber, Max. 1958. *The Rational and Social Foundations of Music*. Trans. G. N. Don Martindale and J. Riedel. Carbondale, IL: Southern Illinois University Press.

Weber, Max. 1972. *Die rationalen und soziologischen Grundlagen der Musik*. Tübingen: Mohr.

Webern, Anton. 1963. *The Path to the New Music*. Ed. Willi Reich, trans. L. Black. Bryn Mawr, PA: Theodore Pressler.

Wegener, Auguste. 1861. Mémoire sur la symphonie des anciens. *Académie royale de Belgique: Extraite des Bulletins*, 2nd series 12 (9–10): 1–8.

Wegman, Rob C. 1995. Sense and sensibility in late-medieval music: Thoughts on aesthetics and "authenticity." *Early Music* 23:299–312.

Wegman, Rob C. 2005. *The Crisis of Music in Early Modern Europe, 1470–1530*. New York: Routledge.

Wegman, Rob C. 2014. What is counterpoint? In *Improvising Early Music*, ed. Rob C. Wegman, Johannes Menke, and Peter Schubert, 11:9–68. Leuven, Belgium: Leuven University Press.

Weiss, Piero, and Richard Taruskin, eds. 2008. *Music in the Western World: A History in Documents*. Belmont, CA: Thomson/Schirmer.

Werner, Eric. 1970. *The Sacred Bridge: Liturgical Parallels in Synagogue and Early Church*. New York: Schocken Books.

Wernicke, Carl. 1977. *Wernicke's Works on Aphasia: A Sourcebook and Review*. Trans. G. H. Eggert. The Hague: Mouton.

West, M. L. 1992. *Ancient Greek Music*. Oxford: Clarendon Press.

West, M. L. 1994. The Babylonian musical notation and Hurrian melodic texts. *Music & Letters* 75:161–179.

Westfall, Richard. 1980. *Never at Rest: A Biography of Isaac Newton.* Cambridge: Cambridge University Press.

Whitehead, Alfred North. 1967. *Science and the Modern World.* New York: Free Press.

Wiener, Norbert. 1965. *Cybernetics: Or the Control and Communication in the Animal and the Machine.* 2nd ed. Cambridge, MA: MIT Press.

Williams, Peter. 1983. The snares and delusions of musical rhetoric: Some examples from recent writings on J. S. Bach. In *Alte Musik, Praxis und Reflexion: Sonderband der Reihe Basler Jahrbuch für Historische Musikpraxis zum 50. Jubiläum der Schola Cantorum Basiliensis*, ed. Peter Reidemeister and Veronika Gutmann, 230–240. Winterthur: Amadeus.

Winternitz, Emanuel. 1982. *Leonardo da Vinci as a Musician.* New Haven, CT: Yale University Press.

Wolff, Christoph. 1991. *Bach: Essays on His Life and Music.* Cambridge, MA: Harvard University Press.

Wolff, Christoph. 2000. *Johann Sebastian Bach: The Learned Musician.* New York: W. W. Norton.

Wolff, Christoph, Arthur Mendel, and Hans T. David, eds. 1999. *The New Bach Reader: A Life of Johann Sebastian Bach in Letters and Documents.* Rev. and enl. ed. New York: W. W. Norton.

World's Fastest Drummer: Tom Grosset Strikes Drum Over 1,200 Times In 60 Seconds. 2016. *International Science Times.* Accessed October 24. http://www.isciencetimes.com/articles/5674/20130718/worlds-fastest-drummer-tom-grosset-strikes-drum.htm.

Wright, Craig M. 1989. *Music and Ceremony at Notre Dame of Paris, 500–1550.* Cambridge: Cambridge University Press.

Yamanoue, Tokiko. 1996. Artificial "attention" in an oscillatory neural network. In *Toward a Science of Consciousness: The First Tucson Discussions and Debates*, ed. Stuart R. Hameroff, Alfred W. Kaszniak, and Alwyn C. Scott, 377–382. Cambridge, MA: MIT Press.

Young, Robert M. 1970. *Mind, Brain and Adaptation in the Nineteenth Century: Cerebral Localization and Its Biological Context from Gall to Ferrier.* Oxford: Clarendon Press.

Yuste, Rafael. 2015. From the neuron doctrine to neural networks. *Nature Reviews. Neuroscience* 16 (8): 487–497.

Zarlino, Gioseffo. 1968. *The Art of Counterpoint: Part Three of Le Istituzioni harmoniche, 1558.* New Haven, CT: Yale University Press.

Zarlino, Gioseffo. 2011. *Le Istituzioni armoniche.* Treviso: Diastema.

Zoubov, V. P. 1961. Nicole Oresme et la musique. *Mediaeval and Renaissance Studies* 5:96–107.

Illustration Credits

Permission for the use of the figures has kindly been given by the following:

Simha Arom (fig. 1.3); Bibliothèque nationale de France, fr. 1665, f. 7r (fig. 4.1); *The Rite of Spring* by Igor Stravinsky © 1912, 1921 by Hawkes & Son (London) Ltd., international copyright secured, all rights reserved, reprinted by permission of Boosey & Hawkes, Inc. (figs. 0.1, 0.2); The British Library/Science Source (fig. 3.5a); The British Museum (fig. 1.6); György Buzsáki (figs. 15.1, 15.2); The Cleveland Museum (fig. 5.3); copyright © by C. F. Peters, used by permission, all rights reserved (fig. 11.10); © 1961 and 1963 by Henmar Press, Inc., used by permission of C. F. Peters Corp., all rights reserved (figs. 12.4, 12.6); reproduced with the permission of the CBC Still Photo Collection, the Estate of Glenn Gould, and Glenn Gould Limited (fig. 12.3a); Harvard University Archives, UAV 605.270.1, Box 8, SC178 (fig. 12.5); Houghton Library, Harvard University, Inc. 6316.10 (A) (fig. 5.1); Johnna Arnold, jka.photo (fig. 7.5b); Josquin Research Project (www.josquin.stanford.edu), Stanford University (figs. 6.10–6.12); Pierre Gouin, Les Éditions Outremontaises © 2008 (figs. 8.2, 10.7); The Metropolitan Museum of Art (www.metmuseum.org), Robert Lehman Collection, 1975 (fig. 3.1), Fletcher Fund, 1956 (7.5a), Rogers Fund, 1917 (fig. 8.5); Joseph Jordania (fig. 1.1); Philip Legge © 2008 (figs. 7.1, 7.2, 7.3, 7.4); Nicole McNeil (fig. 14.4); courtesy of The Master and Fellows of St. John's College, Cambridge, MS 1.15, fol. 144r (fig. 6.3); Ministero dei beni e delle attività culturali e del turismo (fig. 3.8); Museo nazionale della scienza e della tecnologia Leonardo da Vinci, Milan fig. 3.9); National Museum of Denmark (fig. 1.4a); courtesy of the Cajal Institute-CSIC, Madrid, Spain, © CSIC (fig. 14.1); Nino Tsishivili (fig. 1.2); used by permission of European American Music Distribution Company, sole US and Canadian agents for Schott Music, Mainz, Germany, publisher and copyright owner (figs. 11.14, 11.15); St. Gallen, Stiftsbibliothek, Cod. Sang. 390, p. 13 (fig. 2.1); © 1914, 2010 by Universal Edition A.G. Wien UE 35555 www.universaledition.com (fig. 11.6): Universitätsbibliothek Heidelberg, Biblioteca Apostolica Vaticana, Pal. lat. 1342 (figs. 2.3a,b); Warburg Institute Photographic Collection (figs. 1.8, 8.6); Yale University Press, publisher of Emanuel Winternitz, *Leonardo da Vinci as a Musician* © 1982, fig. 12.1, p. 220 (fig. 5.3); Michaël Zugaro (fig. 15.3).

Acknowledgments

My heartfelt thanks to Katie Helke, whose editorial insight and support has meant so much to me. This book owes a lot to her collaboration and enthusiasm, along with her colleagues Judy Feldmann, Gita Manaktala, Margarita Encomienda, and many others at the MIT Press.

This book began and grew through my involvement in the music program of St. John's College in Santa Fe, where I discussed the nature and meaning of polyphony with my fellow students and colleagues over the years. I particularly thank Philip LeCuyer for the conversations beginning thirty-five years ago in which he helped me to start thinking about these matters, particularly the meaning of Thomas Aquinas's views. Phil's profound thoughtfulness and gifts as a thinker remain touchstones for me, as well as his personal generosity and friendship. I also remember with special gratitude the late Bruce Venable, whose deep knowledge of theology (and feeling for music) inspired and instructed me.

I thank Laura Cooley, Jennifer Sprague, and the staff of the Meem Library at St. John's College, who over the years have worked so hard to bring books from distant libraries to Santa Fe. Without their help and the interlibrary loan program, I could not have written this or any of my books.

I thank Elaine Scarry and Anna Henchman for the invitation to present an early form of this work to their seminar at Harvard University on Cognitive Theory and the Arts. I am also grateful to John McCarthy for inviting me to share this work with the School of Philosophy at the Catholic University of America.

Over the years, I have been greatly helped by those who read drafts and generously gave comments and advice, especially Thomas Mathiesen, Mary-Louise Göllner, James Haar, Richard Crocker, and the late Michel Huglo, Their musicological learning and expertise helped me to see my way forward. Angèle Christin kindly read and commented on my use of sociology; Alexei Pesic provided essential guidance in preparing the e-book version. Ssu Weng gave me her thoughts about the neuroscience sections, as did György Buzsáki, to whom I owe special thanks for so generously sharing his expertise. His comments gave me essential guidance on contemporary neuroscience. Andrei Pesic read the whole work and gave me very astute and helpful advice, as did the four anonymous readers whose extensive comments and criticisms were essential in my revisions. The remaining errors are, of course, my responsibility.

The puzzle canon in the dedication spells out the names of Ssu, Andrei, Alexei, and Angèle, the polyphonic minds who inspire and sustain me with their love.

Index